T0140813

Constructive Approaches
to Submanifold Stabilization

Von der Fakultät Konstruktions-, Produktions- und Fahrzeugtechnik
und dem Stuttgart Research Centre for Simulation Technology
der Universität Stuttgart zur Erlangung der Würde eines
Doktor-Ingenieurs (Dr.-Ing.) genehmigte Abhandlung

Vorgelegt von

Jan Maximilian Montenbruck

aus Oberhausen

Hauptberichter: Prof. Dr.-Ing. Frank Allgöwer
Mitberichter: Prof. Murat Arcak, Ph.D.
Prof. Antonis Papachristodoulou, Ph.D.

Tag der mündlichen Prüfung: 17.05.2016

Institut für Systemtheorie und Regelungstechnik
Universität Stuttgart
2016

Bibliografische Information der Deutschen Nationalbibliothek

Die Deutsche Nationalbibliothek verzeichnet diese Publikation in der
Deutschen Nationalbibliografie; detaillierte bibliografische Daten sind
im Internet über http://dnb.d-nb.de abrufbar.

D93

ISBN 978-3-8325-4287-0

Logos Verlag Berlin GmbH
Comeniushof, Gubener Str. 47,
10243 Berlin
Tel.: +49 (0)30 42 85 10 90
Fax: +49 (0)30 42 85 10 92
INTERNET: http://www.logos-verlag.de

To Helen

When I lay asleep, then did a sheep eat at the ivy-wreath on my head,–it ate, and said thereby: "Zarathustra is no longer a scholar." It said this, and went away clumsily and proudly. A child told it to me. I like to lie here where the children play, beside the ruined wall, among thistles and red poppies. A scholar am I still to the children, and also to the thistles and red poppies. Innocent are they, even in their wickedness. But to the sheep I am no longer a scholar: so willeth my lot–blessings upon it! For this is the truth: I have departed from the house of the scholars, and the door have I also slammed behind me. Too long did my soul sit hungry at their table: not like them have I got the knack of investigating, as the knack of nut-cracking. Freedom do I love, and the air over fresh soil; rather would I sleep on ox-skins than on their honours and dignities. I am too hot and scorched with mine own thought: often is it ready to take away my breath. Then have I to go into the open air, and away from all dusty rooms. But they sit cool in the cool shade: they want in everything to be merely spectators, and they avoid sitting where the sun burneth on the steps. Like those who stand in the street and gape at the passers-by: thus do they also wait, and gape at the thoughts which others have thought. Should one lay hold of them, then do they raise a dust like flour-sacks, and involuntarily: but who would divine that their dust came from corn, and from the yellow delight of the summer fields? When they give themselves out as wise, then do their petty sayings and truths chill me: in their wisdom there is often an odour as if it came from the swamp; and verily, I have even heard the frog croak in it! Clever are they–they have dexterous fingers: what doth my simplicity pretend to beside their multiplicity! All threading and knitting and weaving do their fingers understand: thus do they make the hose of the spirit! Good clockworks are they: only be careful to wind them up properly! Then do they indicate the hour without mistake, and make a modest noise thereby. Like millstones do they work, and like pestles: throw only seed-corn unto them!–they know well how to grind corn small, and make white dust out of it. They keep a sharp eye on one another, and do not trust each other the best. Ingenious in little artifices, they wait for those whose knowledge walketh on lame feet,–like spiders do they wait. I saw them always prepare their poison with precaution; and always did they put glass gloves on their fingers in doing so. They also know how to play with false dice; and so eagerly did I find them playing, that they perspired thereby. We are alien to each other, and their virtues are even more repugnant to my taste than their falsehoods and false dice. And when I lived with them, then did I live above them. Therefore did they take a dislike to me. They want to hear nothing of any one walking above their heads; and so they put wood and earth and rubbish betwixt me and their heads. Thus did they deafen the sound of my tread: and least have I hitherto been heard by the most learned. All mankind's faults and weaknesses did they put betwixt themselves and me:–they call it "false ceiling" in their houses. But nevertheless I walk with my thoughts above their heads; and even should I walk on mine own errors, still would I be above them and their heads. For men are not equal: so speaketh justice. And what I will, they may not will!– Thus spake Zarathustra.

— F. Nietzsche, Thus Spake Zarathustra, Chapter XXXVIII: Scholars

Acknowledgements

As a doctoral student, I was advised by Frank Allgöwer, who gave me the freedom to pursue my own ideas by providing the most fertile ground for research of this kind. I am thankful for having had this opportunity. My gratitude extends to the other members of my thesis committee: Murat Arcak, Antonis Papachristodoulou, and Peter Eberhard. Having mentioned Murat, I emphasize that he was the greatest host when I visited him and his group and that he had a major influence on my research. During my time as a doctoral student, I also visited the groups of and was hosted by Rodolphe Sepulchre and Daniel Zelazo, respectively, both of whom made these stays fruitful and unique. During the former of these two stays, Fulvio Forni and Cyrus Mostajeran treated me with their hospitality. I thank Roger Brockett for having initiated this research when we met in St. Petersburg.

While having been an undergraduate student, Andrés Kecskeméthy had the most significant and long-lasting influence on me. Similarly, when I was a doctoral student, the lectures of Peter Lesky altered my way of thinking. At the Institute for Systems Theory and Automatic Control, Mathias Bürger and Gerd Schmidt taught me the basics of control theory. Alike, many scientific subtleties were elaborated with Florian Brunner and Shen Zeng. Florian Brunner was also my office neighbor and I enjoyed our discussions about politics, literature, history, art, and philosophy.

During my time as a doctoral student, I collaborated with Murat Arcak, Mathias Bürger, Hans-Bernd Dürr, Wolfgang Halter, Yuyi Liu, Gerd Schmidt, Georg Seyboth, and Daniel Zelazo. I enjoyed all of these collaborations and, in particular, they revealed to me that research can actually be fun. I am indebted to Florian Bayer, Rainer Blind, Florian Brunner, Christian Ebenbauer, Christian Feller, Matthias Lorenzen, Simon Michalowsky, Matthias Müller, Georg Seyboth, and Shen Zeng for proof-reading this thesis.

My deepest gratitude shall be expressed to my family for their unconditional support. Helen and I have been on this journey together and she contributed to this thesis as much as I did, though in a distinct fashion. Our son Theo brightened my life, even on cloudy days.

Stuttgart, May 2016
Jan Maximilian Montenbruck

Table of Contents

Abstract

We analyze control problems in which a quantity of interest must be brought towards a submanifold of the space in which it evolves. These problems arise in applications such as synchronization, pattern generation, or path following. In our approaches, we focus on constructive solutions to such problems. First, we attempt to bring the controlled system to the form of a gradient system with drift. Under the assumption that the scalar field defining the gradient vector field has strongly convex restrictions to the affinely translated normal spaces of the submanifold, we find that the gradient vector field can be scaled so as to ensure asymptotic stability of the submanifold. This condition further leads to an algebraic characterization of asymptotic stabilizability, expressed in terms of a Riemannian metric. The latter is of particular interest for systems evolving on Riemannian manifolds. As the aforementioned controls usually consume large amounts of energy, we next concentrate on optimal submanifold stabilization in the sense that we ask to bring a quantity towards a desired submanifold whilst, at the same time, maintaining control energy small. It turns out that optimal controls, in this context, are necessarily structured in such a way that they are state feedbacks which depend linearly on a function that maps to the normal spaces of the submanifold while the matrix describing this linear relationship contains the tangent spaces of the submanifold in its nullspace. Under additional assumptions, particular structured controls are also sufficient for optimality. Last, we solely require control energy and integral distance of a quantity to the submanifold to remain finite. To this end, we formulate a concept for input-output considerations of submanifold stabilization and therein exploit small gains, conicity, and passivity for controller design. Throughout, we illustrate our results by various examples.

Deutsche Kurzzusammenfassung

Konstruktive Ansätze zur Stabilisierung von Untermannigfaltigkeiten

Wir analysieren Regelungsaufgaben, innerhalb derer eine Größe, die von Interesse ist, zu einer Untermannigfaltigkeit desjenigen Raumes gebracht werden soll, innerhalb dessen sie verläuft. Diese Probleme tauchen in Anwendungen wie Synchronisation, Mustergenerierung oder Bahnfolge auf. In unseren Ansätzen konzentrieren wir uns auf konstruktive Lösungen solcher Probleme. Zunächst versuchen wir, das geregelte System auf die Form eines Gradientensystems mit Abweichung zu bringen. Unter der Annahme, dass das Skalarfeld, welches das Gradientenvektorfeld definiert, stark konvexe Beschränkungen auf die affin verschobenen Normalräume der Untermannigfaltigkeit hat, stellen wir fest, dass das Gradientenvektorfeld derart skaliert werden kann, dass asymptotische Stabilität der Untermannigfaltigkeit gewährleistet wird. Diese Bedingung führt darüber hinaus zu einer algebraischen Charakterisierung asymptotischer Stabilisierbarkeit, die in Form einer Riemannschen Metrik ausgedrückt wird. Letzteres ist von besonderem Interesse für Systeme, die auf Riemannschen Mannigfaltigkeiten verlaufen. Da die besagten Regelungen für gewöhnlich große Energiemengen konsumieren, konzentrieren wir uns als nächstes auf optimale Stabilisierung von Untermannigfaltigkeiten in dem Sinne, dass wir verlangen, dass eine Größe zu einer Untermannigfaltigkeit gebracht werden soll, während die Stellenergie gleichzeitig klein gehalten werden soll. Es stellt sich heraus, dass Optimalsteuerungen in diesem Kontext notwendigerweise derart strukturiert sind, dass sie Zustandsrückführungen sind, die linear von einer Funktion abhängen, die in die Normalräume der Untermannigfaltigkeit abbildet, während die Matrix, welche diesen linearen Zusammenhang beschreibt, die Tangentialräume der Untermannigfaltigkeit in ihrem Kern enthält. Unter zusätzlichen Annahmen sind spezielle strukturierte Regelungen zudem hinreichend für Optimalität. Zuletzt verlangen wir nur noch, dass Regelenergie und integrierte Distanz einer Größe zur Untermannigfaltigkeit endlich sein sollen. Hierzu formulieren wir ein Konzept zur Ein-Ausgangs Betrachtung der Stabilisierung von Untermannigfaltigkeiten, innerhalb dessen wir kleine Verstärkungen, Konizität und Passivität zum Entwurf von Reglern ausnutzen. Wir veranschaulichen unsere Resultate durchgehend anhand verschiedener Beispiele.

1 Introduction

Control theory deals with problems of enforcing a given behavior in a system. Usually, this amounts to steering a quantity of interest, x, describing the state of the system, to a certain operating condition. If this operating condition is a point among all possible states the system may attain (its state space), then the control problem is called a setpoint stabilization (or setpoint regulation) problem. If, in contrast, the desired operating condition is characterized by a collection of points, and if this collection of points happens to be a submanifold, S, of the state space, most methods which are tailored to solve setpoint stabilization problems are no longer applicable. In analogy to setpoint stabilization, we call such control problems submanifold stabilization problems and herein aim for constructive solutions to these problems.

The essential distinction between setpoint stabilization and submanifold stabilization is depicted in Figure 1.1. The black curves depict the evolution of the quantity x under the controlled system and the arrows indicate forward direction of time. On the left-hand side, all evolutions of the system approach a desired operating condition described by a setpoint (indicated as a blue circle). On the right-hand side, all evolutions of the system approach a desired operating condition described by a submanifold S (depicted as a blue curve).

The interest in such problems stems from applications and is threefold: either a certain nonstationary behavior – the system evolving freely on S – of a system is (i) actually desired, (ii) more practical or (iii) for some reason inevitable.

The former, (i), is the case in path following: movement of the system along a given curve, cf. [68]; pattern generation: controlling a periodic locomotion of a system, cf. [20]; or synchronization: convergence of states of multiple systems towards each other, cf. [46].

Submanifold stabilization can be, (ii), more practical than setpoint stabilization when

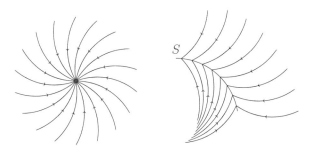

Figure 1.1: Distinction between setpoint stabilization and submanifold stabilization

a controller which stabilizes a submanifold is, in a certain sense, cheaper than a controller which stabilizes a setpoint, for instance when the submanifold can be stabilized by a controller meeting certain communication requirements, which is done in large-scale systems: e.g when communicating and regulating aggregate data, cf. [3]; wind farms: e.g. when the sum of power outputs is regulated via broadcast control, cf. [60]; or vehicle formations: e.g. when relative information is communicated for cooperative behavior, cf. [37].

Stabilization of a submanifold may, (iii), become inevitable when the system is, e.g., underactuated: say a robot with fewer actuators than degrees of freedom, cf. [75]; subject to more velocity constraints than position constraints: say a vehicle which cannot move sidewards but maneuver sidewards by repeatedly moving forwards and backwards, cf. [13]; equipped with too few sensors: say a system whose state can not be uniquely reconstructed from the given measurements and one must opt to only drive a measurement to zero, cf. [10]; or subject to exogenous perturbations: for instance a disturbance from which one aims to decouple a measurement, cf. [92].

1.1 Examples for Submanifold Stabilization Problems

Among other problems, submanifold stabilization problems include setpoint stabilization, path following, pattern generation, synchronization, and control of the center of mass. In the following, we briefly discuss how these problems read as submanifold stabilization problems. The latter four of these five problems are sketched in Figure 1.2 together with the related submanifolds.

The content of the figure is arranged as a table, the left column of which contains illustrations of the different control problems, whilst the right column depicts the submanifolds which are associated with these control problems. The submanifolds are colored blue. Within the figure, for simplicity, we assume that the quantity x whose evolution describes the system under investigation, lives in $\mathbb{R}^2$ and we denote its first element by x_1 and its second element by x_2.

Setpoint Stabilization

The aforementioned setpoint stabilization problem is a submanifold stabilization problem with the submanifold being a singleton, i.e., denoting the setpoint by x_{ref}, the associated submanifold S is $\{x_{\mathrm{ref}}\}$, as x being in S is equivalent to x being equal to x_{ref} (as it is depicted (left) in Figure 1.1, the setpoint stabilization problem is not illustrated in Figure 1.2).

Path Following

In path following, one asks for a system to move along a curve, say $\gamma : \mathbb{R} \to \mathbb{R}^n$ for $\mathbb{R}^n$ being the state space of the system (cf. [68]). This is cast as a submanifold stabilization problem by choosing the submanifold S to be $\gamma(\mathbb{R})$, the image of the curve (cf. Figure 1.2), viz. x being in S is equivalent to existence of some real s such that $\gamma(s)$ is x. The first row in Figure 1.2 is devoted to path following. On the left, we see how the evolution of x, colored blue, approaches a curve, colored black, from some initial condition x_0. On the right, the image of the curve is labeled as the submanifold S.

Pattern Generators

Pattern generators are (neural) circuits which govern locomotion. In artificial pattern generators, one chooses a homotopy circle and associates a certain part of a locomotion with every portion of this circle (cf. [20] or [51]). Consequently, the desired locomotion translates into a periodic evolution on the homotopy circle (cf. Figure 1.2). For instance, one could choose the unit circle $\{x \in \mathbb{R}^2 | \|x\| = 1\}$ for $\mathbb{R}^2$ being the state space of the system and assign the extension of four limbs to four distinct points on the circle. Subsequently, one must fix a direction on the circle and smoothly interpolate between the extension of each two consecutive limbs. In the second row of Figure 1.2, the pattern generator problem is illustrated. On the left, we see how a locomotion pattern, illustrated by blue circles, is associated to a circle, on which a direction is fixed, illustrated by arrows. On the right, we see how the circle is embedded into $\mathbb{R}^2$ as a submanifold. If the locomotion pattern is associated to the circle in $\mathbb{R}^2$, then submanifold stabilization translates to pattern generation.

Synchronization

In synchronization problems, one considers a collection of n systems and asks for their convergence towards each other (cf. [46]). Letting $\mathbb{R}$ be the state space of each of these systems, the submanifold S associated with the synchronization problem is the diagonal span $(\{1_n\})$ with 1_n being the n-fold vector of ones (cf. Figure 1.2). The reason for this is that having $x_i = x_j$ for all i and j is equivalent to $(x_1, \ldots, x_n)$ lying in the span of 1_n, in particular $x = x_i 1_n$ for any i. The synchronization problem is illustrated in the third row of Figure 1.2. On the left, we see how the evolutions of two systems, colored blue, approach each other as time increases. On the right we see the diagonal of $\mathbb{R}^2$.

Controlling the Center of Mass

If one wants to move a population of n systems as a whole, then this amounts to controlling the center of mass (arithmetic mean) of the population (cf. [60]). Thus, the submanifold describing the desired operating condition is precisely the orthogonal complement of span $(\{1_n\})$ in $\mathbb{R}^n$, the antidiagonal (cf. Figure 1.2). This is due to the fact that the center of mass is zero if and only if $(x_1, \ldots, x_n)$ is contained in the antidiagonal, for instance (x_1, x_2) is in S if and only if x_1 is $-x_2$. The last row in Figure 1.2 illustrates this class of control problems. On the left, we see how some initial distribution of states, colored blue, which has nonzero center of mass, eventually reaches a distribution with zero center of mass (the forward direction of time is indicated by the arrows). On the right, the antidiagonal of $\mathbb{R}^2$ is depicted.

We later also encounter other (more complex) submanifold stabilization problems, one of which is the formation control problem (wherein one wants a collection of systems to attain a configuration of a given shape, cf. [25]) and one of which is the satellite surveillance problem (wherein one wants a reconnaissance satellite to face a point on earth, say with a telescope, cf. [26]).

1.2 Related Work

The submanifold stabilization problem as such has so far not been addressed, except for the topological obstructions to submanifold stabilization, i.e. the necessary conditions for solving submanifold stabilization problems, studied by Mansouri [61, 62]. Yet, these conditions are, by their very nature, neither constructive nor applicable to controller synthesis.

Nevertheless, other results are related to submanifold stabilization directly or indirectly. Most earlier results are of purely analytic nature whilst some more recent results also contain constructive elements. We first discuss the former to then discuss the latter.

Analytic Approaches

As for analytic approaches, the possibly foremost important result for attractivity of invariant sets is LaSalle's invariance principle [57], which provides sufficient conditions for solutions of a differential equation to approach an invariant set from all initial conditions contained in a compact invariant neighborhood. A general stability theory for invariant sets, extending Lyapunov's direct method to necessary and sufficient conditions for an invariant set to be (asymptotically) stable, was presented by Zubov [100] and by Bhatia and Szegő [8]. Stability properties of a single solution are determined by its Lyapunov exponents and can be investigated via linearization (cf. [44, section 5.8] and references therein; the original reference appears to be [59]). An extension of the Hartman-Grobman theorem, i.e. that a flow is conjugate to its linearization in the neighborhood of a hyperbolic equilibrium, to normally hyperbolic invariant submanifolds, was derived by Pugh and Shub [72]. For stability of subspaces, a framework called partial stability was established by Vorotnikov [88]. Stability of periodic orbits can be investigated via its Floquet multipliers after linearization, or via its Poincaré map, which is a (discrete) mapping describing how solutions evolve on a transversal section of the periodic orbit (cf. [1, section 23]; the original references appear to be [38] and [71], respectively).

Constructive Approaches

As for constructive approaches, the most important technique is the transverse feedback linearization of Nielsen and Maggiore [69], wherein one computes controllability coefficients in transverse directions to an invariant submanifold. For stability of general closed sets, backstepping [35] and passivity [34] techniques were proposed by El-Hawwary and Maggiore, both with focus on dimension reduction techniques. The aforementioned transverse feedback linearization technique was originally applied to periodic orbits by Banaszuk and Hauser [7], for which frequency domain stability criteria were derived by Yakubovich [96] and a heuristic delayed feedback was proposed by Pyragas [73]. For stabilizability of subspaces, the controllability subspaces of Wonham and Morse are essential [95], which were further investigated within the output stabilization problem of

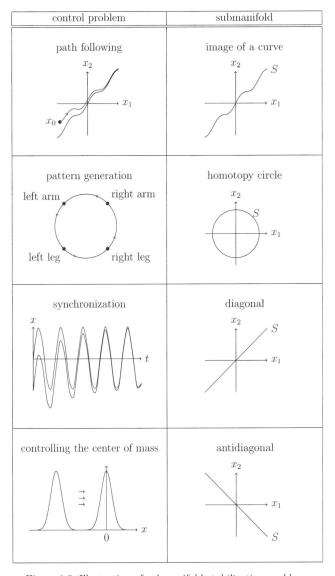

Figure 1.2: Illustration of submanifold stabilization problems

Wonham [94, Chapter 4] (or [10]), that is, in turn, a special case of the output regulation problem of Francis [41]. In the nonlinear case, the output regulation framework of Isidori and Byrnes [49] allows to render rather general families of solutions attractive.

1.3 Contribution and Structure of the Thesis

We study submanifold stabilization problems. In our efforts to solve these problems, we derive constructive results in the sense that we provide approaches for how to – mostly graphically or through structural properties of the control – explicitly construct control laws that stabilize the given submanifold. This brings us into the position to provide particular control laws for certain problems, as we illustrate on several examples.

Chapter 2

In chapter 2, we discuss asymptotic stabilization of a submanifold S for systems described by differential equations of the form

$$\dot{x} = f(x) + \sum_{i=1}^{m} g_i(x) u_i(x) \tag{1.1}$$

with u_i the controls sought. Therein, section 2.1 is devoted to the case where the state space of (1.1) is $\mathbb{R}^n$ (i.e. S is a submanifold of $\mathbb{R}^n$) whilst section 2.2 is devoted to the case where the state space of (1.1) is itself a Riemannian manifold (i.e. S is a submanifold of a Riemannian manifold). Throughout the chapter, i.e. in both sections, our approach is to bring the closed loop to the form of a gradient system with drift. In section 2.1 we show that, should the scalar field defining the gradient vector field be strongly convex when restricted to the (affinely translated) normal spaces of the submanifold, then the gradient vector field can be scaled so as to guarantee asymptotic stability of the submanifold. Further, the rate at which solutions of (1.1) approach S can be modulated ad libitum. Should the given submanifold not be an invariant set, then we show how one can still render an arbitrarily small neighborhood of the submanifold asymptotically stable. Section 2.2 extends these results to the case in which solutions of (1.1) evolve in some Riemannian manifold. Thereafter, we focus on the question under which circumstances it is possible to bring (1.1) to the form of a gradient system with drift. We present an algebraic characterization with which we can answer this question in the affirmative or in the negative. Should the submanifold be an equivalence class for a given equivalence relation, then the aforementioned algebraic characterization can be interpreted in terms of horizontal and vertical spaces. The main results of section 2.1 were published in [M14] and the main results of section 2.2 were published in [M3].

Chapter 3

In chapter 3, we discuss optimal stabilization of submanifolds, i.e. how functionals involving the integral

$$\int d(x(t), S)^2 \, dt, \tag{1.2}$$

plus some functional quantifying control energy, can be infimized (minimized if possible) by appropriate choice of the control, where $d(x, S)$ refers to the infimal Euclidean distance of x to all points in S. We show that all such optimal controls are necessarily structured state feedbacks in the following sense: optimal controls are necessarily state feedback which linearly depend on a function mapping to the normal spaces of S whilst the matrix defining this linear relationship has the tangent spaces of S contained in its nullspace. Under the additional assumption that S is an invariant set, we explicitly construct one such optimal control, thus revealing that structured controls are also sufficient for optimality. Should all involved vector fields be constant when represented in a basis of the normal spaces of S (plus a suitable controllability assumption), then we are able to compute the aforementioned optimal control from an algebraic Riccati equation. The main results of this chapter were published in [M4].

Chapter 4

In chapter 4, we discuss input-output approaches to submanifold stabilization, i.e. how systems described by relations of the form

$$(u, x) \in H_1 \tag{1.3}$$

on sets of Lebesgue measurable functions with certain finite integrals can be coupled with another such relation (the controller) such as to ensure that the function x which is the output of H_1 eventually decays to a given submanifold S. We recover generalizations of the small-gain theorem, the feedback theorem for conic relations, and the feedback theorem for passive systems. Our passivity notion resembles a persistence-of-excitation condition: persistently exciting control signals in the normal spaces of S turn out to be necessary for input strict passivity. In order to facilitate application of our feedback theorem for passive systems, we propose a multiplier technique based upon the Zames-Falb multipliers. The main results of this chapter were published in [M6].

The links between the chapters are as follows: the controllers derived in chapter 2 will usually require large control energy. To cope with this issue, we ask for optimal solutions, letting control energy remain small while bringing solutions towards S, in chapter 3. In chapter 4, we relax these requirements and merely ask for control energy and (integral) distance to S to remain finite, enabling us to broaden our analysis from systems described by differential equations to systems described by input-output relations.

Apart from some general concepts introduced in the beginning of section 2.1, all parts of the thesis are self-contained (with a few exceptions which are explicitly indicated). Notation is introduced in the order of appearance and some terminology is employed without explicitly introducing it.

2 Asymptotic Stabilization of Submanifolds via Gradient Vector Fields

In this chapter, we study (so-called input affine) systems described by differential equations of the form

$$\dot{x} = f(x) + \sum_{i=1}^{m} g_i(x) u_i(x), \qquad (2.1)$$

with f the so-called drift vector field, g_i the so-called control vector fields, and the question of how to find controls u_i such that a given submanifold S of the space in which solutions of (2.1) evolve, its state space, becomes an asymptotically stable invariant set. An invariant set S is said to be stable if, for every neighborhood of the invariant set, there exists a neighborhood with the property that all solutions initialized in the latter remain within the former for all times. It is said to be asymptotically stable if, in addition, there exists a neighborhood from which all solutions approach S. Should S be an asymptotically stable set, we call the set of points from which solutions approach S its region of asymptotic stability (we refer to [8] for these definitions). Although the reachable sets for input affine systems are characterized completely (cf. [47]), it is recognized that asymptotic stabilization by feedback is harder, and, moreover, possible in fewer cases than reachability [19].

We study two cases in this chapter: in section 2.1, we discuss the case where the state space of (2.1) is $\mathbb{R}^n$ whilst in section 2.2, we focus on the (more general) case where the state space of (2.1) is a Riemannian manifold. In both sections, our approach is to bring the closed loop to the form of a gradient system with drift, i.e. to let the vector field

$$x \mapsto g_1(x) u_1(x) + \cdots + g_m(x) u_m(x) \qquad (2.2)$$

be integrable (or, equivalently, have path independent line integrals). Thereupon, we scale the scalar field of which (2.2) is the gradient by a scalar k. Under the assumption that S is an invariant set and under suitable assumptions on the scalar field, we consequently show that there exists some lower bound k_0 with the property that, if we choose k such that it exceeds this bound, S becomes asymptotically stable. The intuition behind this tuning procedure for the control gain k is the following: if one integrates (2.2) and solves for the functions u_i for $k = 1$ in the first place, but one knew that the aforementioned bound k_0 exists, then it is sufficient to consequently perform the scaling $u_i \mapsto k u_i$ for some sufficiently large k in order to stabilize S asymptotically.

9

2.1 Asymptotic Stabilization of Submanifolds of $\mathbb{R}^n$

In this section, we investigate systems of the form (2.1) evolving in $\mathbb{R}^n$, i.e. with drift vector field $f : \mathbb{R}^n \to \mathbb{R}^n$ and control vector fields $g_1, \ldots, g_m : \mathbb{R}^n \to \mathbb{R}^n$ given, the controls $u_1, \ldots, u_m : \mathbb{R}^n \to \mathbb{R}$ yet being sought. We assume that the submanifold S of $\mathbb{R}^n$ is an invariant set of the unforced system

$$\dot{x} = f(x) \tag{2.3}$$

and we assume that S is compact and smoothly embedded. The integrability Ansatz

$$g_1 u_1 + \cdots + g_m u_m = -k\nabla\phi, \tag{2.4}$$

with $\phi : \mathbb{R}^n \to \mathbb{R}$ being a continuously differentiable function and $\nabla\phi : \mathbb{R}^n \to \mathbb{R}^n$ denoting its gradient, will be assumed throughout the chapter even if not explicitly stated, and is meant to improve the conciseness of presentation in the sense that it allows us to formulate properties of the controls solely in terms of properties of ϕ. We will later (specifically in the following section) outline how our results would read if we had not taken this Ansatz. The main results of this section were published in [M14].

We let $T_x S$ denote the tangent space of S at x and $N_x S$ the orthogonal complement of $T_x S$ in $\mathbb{R}^n$. The normal bundle NS is the vector bundle

$$NS = \bigsqcup_{x \in S} N_x S \tag{2.5}$$

composed of the fibers $N_x S$. A neighborhood of S is said to be tubular if it is the diffeomorphic image of $\mu : NS \to \mathbb{R}^n$, $(x, v) \mapsto x + v$ and the tubular neighborhood theorem asserts that such neighborhoods exist for embedded submanifolds. Let $\mathsf{P}_1 : NS \to S$, $(x, v) \mapsto x$ be the bundle projection and, in analogy, define the fiber projection $\mathsf{P}_2 : NS \to \mathbb{R}^n$, $(x, v) \mapsto v$. The map $r := \mathsf{P}_1 \circ \mu^{-1} : U \to S$ is a smooth retraction of the tubular neighborhood onto S and every member of U decomposes uniquely as a coordinate on S, given by r, and a fiber coordinate

$$\pi := \mathsf{P}_2 \circ \mu^{-1} : U \to \mathbb{R}^n \tag{2.6}$$

quantifying the deviation from S, i.e. the decomposition

$$\mathsf{P}_1 \circ \mu^{-1} + \mathsf{P}_2 \circ \mu^{-1} = r + \pi = \mathrm{id}, \tag{2.7}$$

with $\mathrm{id} : U \to U$ being the identity $x \mapsto x$, remains satisfied. Thus, π defines the "natural" Lyapunov function candidate

$$V := \frac{1}{2} \|\pi\|^2 : U \to \mathbb{R} \tag{2.8}$$

which is smooth due to the smoothness of r, vanishing on S, and positive elsewhere. We define the affine fibers $F_x := \mu(x, N_x S)$ resulting from translation of the fibers $N_x S$ via addition of x. The following theorem is our main result of this section and was published in [M14].

Theorem 1. Let U be a tubular neighborhood of S. Let S be a critical submanifold of ϕ and let ϕ be regular on $U \setminus S$. If f is continuous as well as locally one-sided Lipschitz continuous and, for all $x \in S$, the restriction of ϕ to the intersection of U with F_x is strongly convex, then, for every $\alpha > 0$ such that $V^{-1}(\{\alpha\})$ is contained in U, there exists a $k_0 \geq 0$ such that, for all $k > k_0$, S is an asymptotically stable invariant set of (2.1) and $V^{-1}([0, \alpha])$ is a subset of its region of asymptotic stability.

Proof. We first compute the gradient vector field $\nabla V : U \rightarrow \mathbb{R}^n$ of V in order to evaluate the Lie derivative of V along $f - k\nabla\phi$.

First note that the gradient of V has the form

$$\nabla V : x \mapsto (I_n - \mathrm{J}_r(x)) \, \pi(x) \tag{2.9}$$

with $\mathrm{J}_r : U \rightarrow \mathbb{R}^{n \times n}$ denoting the Jacobian of r and I_n being the $n \times n$ identity matrix. Next, note that the preimage of $\{x\}$ under r is the normal space $N_x S$ translated by addition with x, i.e. the affine space $\{x\} + N_x S = F_x$, whose tangent space is precisely the normal space of S at x. This lets us conclude that the Jacobian J_r of r contains the normal spaces of S in its nullspaces, viz. $\ker(\mathrm{J}_r(x)) \supset N_{r(x)} S$. As r retracts along normal spaces of S, we have that π maps x to a member of $N_{r(x)} S$ and hence that ∇V is just π.

Next, compute the Lie derivative $L_{f-k\nabla\phi} V : U \rightarrow \mathbb{R}$ of V along $f - k\nabla\phi$. Due to linearity, the Lie derivative splits as the sum

$$L_{f-k\nabla\phi} V = L_f V - k L_{\nabla\phi} V \tag{2.10}$$

and we investigate the Lie derivatives $L_f V$ of V along f and $L_{\nabla\phi} V$ of V along $\nabla\phi$ separately.

First, investigate the Lie derivative $L_f V$ of V along f. We have argued that ∇V is given by π, which maps to the normal spaces of S. As S is an invariant set of $\dot{x} = f(x)$, it is necessary that $f \circ r$ maps to the tangent spaces of S and that hence, as the normal spaces are the orthogonal complements of the tangent spaces in $\mathbb{R}^n$, the vector fields ∇V and $f \circ r$ are orthogonal. It follows that the Lie derivative of V along f is precisely the inner product of ∇V and $f - f \circ r$. However, f was assumed to be locally one-sided Lipschitz continuous, i.e. for every x, there exists an open neighborhood U_x of x and a real q_x such that, for all x' in U_x, the overestimate

$$(f(x) - f(x')) \cdot (x - x') \leq q_x \|x - x'\|^2 \tag{2.11}$$

holds. Choose some $\alpha > 0$ such that $V^{-1}(\{\alpha\})$ is contained in U. As S is compact, $V^{-1}([0, \alpha])$ is compact and hence its open cover $\{U_x\}_{x \in U}$ has finite subcover, over whom a maximal q_x exists, which we denote by q. Recalling that $\pi : x \mapsto x - r(x)$ leaves us with the overestimate

$$L_f V \leq 2qV \tag{2.12}$$

for the Lie derivative of V along f, which is valid on $V^{-1}([0, \alpha])$.

Next, investigate the Lie derivative $L_{\nabla\phi} V$ of V along $\nabla\phi$. As we assumed that S is a critical submanifold of ϕ, we have that $(\nabla\phi) \circ r$ sends every element of U to the zero vector. Thus, the Lie derivative of V along $\nabla\phi$ is precisely the inner product of

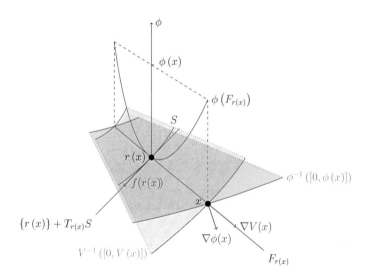

Figure 2.1: Key concepts from the proof of Theorem 1

∇V and $\nabla \phi - (\nabla \phi) \circ r$. As, for all $x \in S$, the restriction of ϕ to the intersection of U with F_x is strongly convex, we have that for every $x \in S$, there exists a $\lambda_x > 0$, varying continuously with x, such that, for all x' in $U \cap F_x$,

$$(x' - x) \cdot (\nabla \phi (x') - \nabla \phi (x)) \geq \lambda_x \|x - x'\|^2 \qquad (2.13)$$

holds true. Due to compactness of S and continuity of $x \mapsto \lambda_x$, some minimal λ_x exists, which we denote by λ. Again recalling that $\pi : x \mapsto x - r(x)$ yields the underestimate

$$L_{\nabla \phi} V \geq 2\lambda V \qquad (2.14)$$

for the Lie derivative of V along $\nabla \phi$.

Combining the overestimates (2.12) for the Lie derivative of V along f and the underestimate (2.14) for the Lie derivative of V along $\nabla \phi$ yields the overestimate

$$L_{f - k \nabla \phi} V \leq 2 (q - k\lambda) V \qquad (2.15)$$

for the Lie derivative of V along $f - k\nabla \phi$ after substitution into (2.10). Setting k_0 to q/λ, we find that, for all $k > k_0$, the Lie derivative of V along $f - k\nabla \phi$ becomes negative for all $x \in V^{-1}([0, \alpha]) \setminus S$. As $V^{-1}([0, \alpha])$ is a neighborhood of S and S is compact, this reveals that S is an asymptotically stable invariant set by Lyapunov's direct method. As, moreover, $L_{f - k \nabla \phi} V$ is nonpositive on $V^{-1}([0, \alpha])$ and the latter is compact, it is a forward invariant set. Hence, by LaSalle's invariance principle, $V^{-1}([0, \alpha])$ is a subset of the region of asymptotic stability of S. This concludes the proof. $\qquad \square$

A few comments are in order. One thing that we owe is a geometric interpretation of the key concepts from the proof, which is provided in Figure 2.1. For a point x within the neighborhood of S that we considered, the sublevel set $V^{-1}([0, V(x)])$, depicted in gray, is a subset of the tubular neighborhood U and thus it is a tubular neighborhood itself: from each of its elements, say x, here indicated by a circle, one can trace back the affine fiber $F_{r(x)}$, plotted as a black line, uniquely until reaching $r(x)$. The boundary $V^{-1}(\{V(x)\})$ of $V^{-1}([0, V(x)])$ consists of equidistant points to S and its intersection with level surfaces $\phi^{-1}(\{\phi(x)\})$ of ϕ can not be orthogonal due to strong convexity of ϕ restricted to the affine spaces $F_{r(x)}$, with the restriction depicted in blue. More, the preimage of $[0, \phi(x)]$ under ϕ (ϕ is constant on S as S is its critical submanifold and we here thus assume that $\phi \circ r : x \mapsto 0$ without loss of generality), here depicted in light blue, must contain $r(x)$ inside its intersection with the sublevel set of V. In particular, the angle enclosed between $\nabla \phi(x)$ and $\nabla V(x)$, both plotted red, is acute, i.e. their inner product is nowhere nonnegative. On the other hand, $f(r(x))$, plotted red as well, is a vector lying in the tangent space of S at $r(x)$ and due to its local one-sided Lipschitz continuity, it can not deviate from this tangent direction arbitrarily fast within a neighborhood of S.

With this graphical interpretation in mind, we find that Theorem 1 is constructive in the sense that it reduces the controller design problem to figuring a function whose restrictions to the affine spaces resulting from translation of the normal spaces of the submanifold are strongly convex. This simplifies the problem as the dimension of these affine spaces is precisely the codimension of S in $\mathbb{R}^n$, i.e. one merely needs to construct a family of functions, the dimension of whose domains are the codimension of S in $\mathbb{R}^n$ and whose codomains have dimension 1. For instance, as S is compact, one can choose a finite ϵ-net of S (which one can explicitly compute, e.g., via farthest-first traversal or Lloyd's algorithm / Voronoi iteration) and construct a strongly convex restriction of ϕ to F_x, the dimension of whose domain is the codimension of S in $\mathbb{R}^n$, for any x in the given ϵ-net. The function ϕ could subsequently be obtained by smoothly interpolating between the different restrictions of ϕ. Alternatively, as the normal spaces of S all result from each other by rotation, one could construct one such restriction, say the restriction of ϕ to F_x, and obtain another restriction, say the restriction of ϕ to $F_{x'}$, through

$$\phi\big|_{F_{x'}} : x'' \mapsto (c \circ r)(x'') \left(\phi\big|_{F_x} \circ \mu\right)(x, \text{rot}\,\pi(x'')) \tag{2.16}$$

wherein c is a scalar field on S which is uniformly bounded from below by a positive scalar and rot is a member of the special orthogonal group $SO(n)$ satisfying $\text{rot}\,N_{x'}S = N_x S$ (such a rotation matrix can be obtained by left-multiplying an orthonormal matrix whose first rows are an orthonormal basis of $N_{x'}S$ with an orthonormal matrix whose first columns are an orthonormal basis of $N_x S$). When transitioning from one restriction to another in this fashion, c introduces an additional scaling of the previous restriction.

The setting of the foregoing theorem appears to be very particular and different applications will raise questions for individual extensions and relaxations, some of which we elaborate in the following: we discuss how the assumptions that $g_1 u_1 + \cdots + g_m u_m$ is integrable, that S is compact, that S is a submanifold, and that S is an invariant set of the unforced system can be relaxed as well as how local one-sided Lipschitz continuity may be replaced by local Lipschitz continuity. Further, we mention how certain convergence rates / performance bounds can be achieved and how the parameter k_0 can be approximated algorithmically.

Integrability

We first discuss the integrability assumption (2.4) on $g_1 u_1 + \cdots + g_m u_m$. Expressed as line integral along some curve $\gamma : [0,1] \rightarrow F_{r(x)}$ with $\gamma(0) = r(x)$, $\gamma(1) = x$, our assumption reads

$$\phi(x) = \int_0^1 \sum_{i=1}^m (g_i u_i \circ \gamma)(s) \cdot \dot{\gamma}(s) \, \mathrm{d}\, s, \qquad (2.17)$$

a fiber integral, with the requirement that the value $\phi(x)$ be independent of the specific choice of γ. If one does not want to explicitly presume integrability, then one could still decompose $g_1 u_1 + \cdots + g_m u_m$ into integrable and nonintegrable parts (via Helmholtz, or, more general, Hodge decomposition) and assign the former to $-k\nabla\phi$ but the latter to f. It is, however, worth stressing that integrability of control vector fields is not unusual (cf. [18] for the direct relation of asymptotic stabilizability to involutivity) and under some circumstances even necessary, e.g. in input strictly passive controllers if one wants to apply Zames-Falb multipliers [76]. On the other hand, if we were not to assume (2.4), we would have to handle the underestimate (2.14) in terms of $g_1 u_1 + \cdots + g_m u_m$ and hence be left with the inequality

$$\pi \cdot \sum_{i=1}^m g_i u_i \leq -\lambda V \qquad (2.18)$$

which, for a given positive λ, has solution in $u_1, \ldots, u_m$ if and only if, for all x in U, there exists an i such that $g_i(x)$ is not in $T_{r(x)}S$. This condition appears feasible, particularly when recalling that the inner product has not yet been fixed. In particular, the foregoing inequality might have solution in $u_1, \ldots, u_m$ under choice of one inner product while it has no solution under choice of the other (whereas asymptotic stability of S is independent of the inner product). We elaborate this condition further in the next section and therein also discuss the situation in which the inner product varies (smoothly) with x, i.e. we turn our attention to Riemannian manifolds.

Example 1. Let S be the unit circle, $\mathbb{S}_1$, embedded in $\mathbb{R}^2$. Consider the system

$$\dot{x} = (\|x\| - 1) x + \Omega x + \frac{1}{\|x\|} x u(x), \quad \Omega := \begin{bmatrix} 0 & -1 \\ 1 & 0 \end{bmatrix} \qquad (2.19)$$

whose drift vector field f and control vector field g,

$$f(x) = (\|x\| - 1) x + \Omega x, \quad g(x) = \frac{1}{\|x\|} x \qquad (2.20)$$

satisfy $(f \circ r)(x) \in T_{r(x)}\mathbb{S}_1$ and $g(x) \in N_{r(x)}\mathbb{S}_1$, respectively. The former implies that $\mathbb{S}_1$ is an invariant set of the unforced system (2.3) whilst the latter implies that (2.18) remains satisfied. The drift vector field f is plotted on the left-hand side of Figure 2.2, depicted as black arrows, from which we can figure that $\mathbb{S}_1$ is an unstable invariant set of the unforced system. Moreover, as f is nonvanishing on $\mathbb{S}_1$ but g acts only normal to

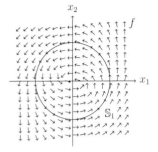
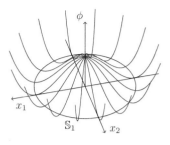

Figure 2.2: Drift vector field (left) and scalar field (right) from Example 1

$\mathbb{S}_1$, asymptotically stabilizing a specific point in $\mathbb{S}_1$ is an impossible task. Now suppose that $u : \mathbb{R}^2 \to \mathbb{R}$ is sought such that $\mathbb{S}_1$ becomes an asymptotically stable invariant set of (2.19). Pursuing the approach proposed above, we first choose a function ϕ whose restriction to any normal space of $\mathbb{S}_1$ is strongly convex, say

$$\phi(x) = \frac{1}{3} \|x\|^3 - \frac{1}{2} \|x\|^2 + \frac{1}{6}, \tag{2.21}$$

the restrictions of which to some normal spaces of $\mathbb{S}_1$ are plotted on the right-hand side of Figure 2.2 in blue. The gradient of ϕ is $x \mapsto (\|x\| - 1) x$, such that the integrability condition (2.4) is solved by $u(x) = -k \|x\| (\|x\| - 1)$. Here, the closed loop system reads

$$\dot{x} = (1 - k)(\|x\| - 1) x + \Omega x, \tag{2.22}$$

for which $\mathbb{S}_1$ becomes asymptotically stable as k exceeds $k_0 = 1$. We notice that the control reverses the normal components of the drift (for $k > k_0$) whilst leaving its tangent components unchanged.

The content of this example was previously published in [M14].

As mentioned in the previous chapter, the interest in asymptotic stabilization of homotopy circles partially stems from artificial pattern generators (again cf. [20] or [51]), in which locomotion sequences (of robots) are first associated to portions of an orbit to then asymptotically stabilize that orbit (and hence the chosen locomotion), usually with an electrical circuit (it is thus advantageous if the orbit is known in closed form, a property which, e.g., the Van der Pol oscillator does not possess). In such applications, robustness, reliability, safety, or energy consumption are crucial to a successful implementation rather than rigorous convergence analysis as pursued above. In a brief experiment, we thus next demonstrate that a (physical) circuit realization of the proposed control $u : x \mapsto -k \|x\| (\|x\| - 1)$ is indeed applicable to a (physical) realization of the system (2.19) and successfully achieves asymptotic stabilization of the circle. This also emphasizes that the suggested control can be realized as a electrical circuit, in contrast to some modern controllers which necessitate implementation on a digital computer. A sketch of the circuit realization of our control is depicted in Figure 2.3.

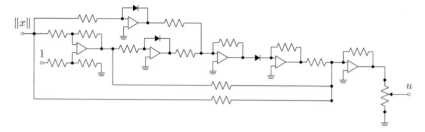

Figure 2.3: Circuit realization of the control $u(x) = -k\,\|x\|\,(\|x\| - 1)$

The realization consists of a multiplier, a summing amplifier, and a potentiometer, taking $\|x\|$ and 1 Volt supply voltage as input potentials, the pins of which are positioned on the far left of the figure (we efficiently realized the system (2.19) as a circuit with output voltage $\|x\|$). The output voltage of the control, u, the pin of which is depicted on the far right, is adjusted via a potentiometer in order to realize our gain k. All amplifiers are operational amplifiers with inverting input on top. The resistors and diodes are arbitrary up to scaling of u, depending on resistance and reverse leakage current, respectively.

With this physical circuit realization of the control $u(x) = -k\,\|x\|\,(\|x\| - 1)$, we performed four exemplary experiments on the physical realization of the system (2.19) (the printed circuit boards were assembled by Fabian Pfitz). The voltages x_1, x_2 were graphed with an oscilloscope during these experiments, photographs of whose traces are depicted in Figure 2.4. The photographs are arranged in such a fashion that k is realized as $1.2 > k_0$ in the right column and as $0.8 < k_0$ in the left column whilst the initial condition of (2.19) was placed inside the unit circle in the upper row and outside the circle in the bottom row. The initial conditions were realized by capacitors and set to $(0.5, 0.5)$ Volts and $(2, 2)$ Volts in the left and right column, respectively. In the upper left photograph, we see how the two voltages asymptotically approach the origin. In the lower left, the voltages asymptotically approach the voltage support of $\|x\|_\infty = 15$ Volt. In the right column, the voltages approach the unit circle, as desired, following a transient phase resulting from capacitor discharging. In summary, all solutions for the case $k < k_0$ are repelled from the circle and all solutions for the case $k > k_0$ are attracted to the circle, letting us conclude that the (theoretical) bound k_0 is sharp enough to be relevant for practical implementations. Furthermore, the experiments verify that the proposed control is robust enough to cope with the uncertainties that are inherent to physical systems (voltage fluctuations, production tolerances, etc.).

Compactness

The compactness of S was employed at several instances of the proof. To consider closed, but unbounded S, one must overcome the finite subcover and continuity arguments in the proof and seek for replacements of Lyapunov's direct method for asymptotic stability of compact invariant sets and LaSalle's invariance principle for compact invariant subsets of the region of asymptotic stability. The former tasks are solved by assuming

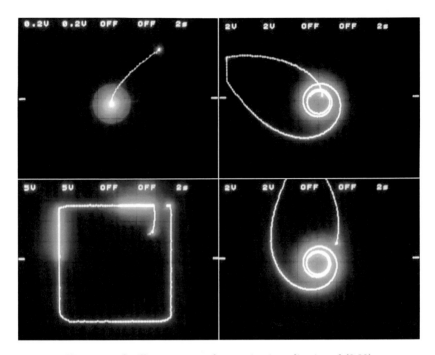

Figure 2.4: Oscilloscope traces from a circuit realization of (2.22)

that the inequalities (2.11) and (2.13) hold uniformly on U and S respectively, i.e. that there exist q such that, for all x in U, $q \geq q_x$ and $\lambda > 0$ such that, for all x in S, $\lambda \leq \lambda_x$. The latter tasks turn out to be more involved. To prove asymptotic stability, we could either presume forward existence of all solutions initialized in a neighborhood of S to then subsequently use the fact that V is over- and underestimated by continuous strictly increasing functions in $x \mapsto d(x, S)$ which are zero at zero (so-called class $\mathcal{K}$ functions, cf. [45]), apply Grönwall's inequality to (2.15), and then invoke the Lyapunov theory for closed sets from [8, chapter V, section 4], or restrict our attention to a closed invariant set U' such that $S \cap U'$ is compact and U' has nonempty intersection with a tubular neighborhood of S to then invoke the relative stability theory from [8, chapter V, section 5] and show that S is asymptotically stable relative to U'. As for the region of asymptotic stability, although one can (implicitly) prove existence of a tubular neighborhood from which solutions approach S [93], providing an underestimate for the region of asymptotic stability of closed but unbounded sets will require to presume that solutions initialized in these sets remain bounded and have no limiting points on their boundary [58, chapter 2, section 6]. In our theorem, these complications are reflected by the fact that the preimages of V contained in tubular neighborhoods of S provide the underestimates for the region of asymptotic stability of S, while noncompact embedded submanifolds may have no tubular neighborhood containing preimages of V. This property is, however, guaranteed for compactly embedded submanifolds, following the construction in [16, chapter II, section 11].

Example 2 (Synchronization). In synchronization problems (cf. [46]), one, for instance, considers a collection of n systems

$$\dot{x}_i = f(x_i) + u_i(x), \quad i = 1, \ldots, n \tag{2.23}$$

with drift vector fields $f : \mathbb{R}^p \to \mathbb{R}^p$ and the controls $u_1, \ldots, u_n : \mathbb{R}^{np} \to \mathbb{R}^p$ sought such that the solutions of the systems converge towards each other, i.e. the stack of solutions $x = (x_1, \ldots, x_n)$ approaches the diagonal $\mathcal{D} = \{(x_1, \ldots, x_n) | x_1 = \cdots = x_n\}$ of the product space $(\mathbb{R}^p)^n$ resulting from taking the n-fold Cartesian product of the state spaces $\mathbb{R}^p$ of the individual systems with itself. Additional requirements on the controls relevant in applications are, e.g., that the solutions of the closed loop contained in $\mathcal{D}$ are the solutions of the unforced systems, or that the controls can be computed based on relative quantities of the form $x_i - x_j$ (cf. [37]). This is motivated from relative sensing mechanisms arising in nature and engineering, such as displacement measurements. Both these additional requirements can be satisfied using the approach proposed above: as $\mathcal{D}$ is a subspace of $\mathbb{R}^{np}$, a quadratic function

$$\phi(x) = \frac{1}{2} x \cdot (P \otimes I_p) x, \tag{2.24}$$

where $\otimes$ denotes the Kronecker product, can be found whose restriction to the orthogonal complement of $\mathcal{D}$ in $\mathbb{R}^n$, and all its affine translations, is strongly convex. This is done by letting P be a symmetric matrix whose nullspace is the span of the vector of ones, but which is positive semidefinite. For instance, letting P be the weighted Laplacian matrix of some undirected, connected, weighted graph with n vertices is a possible Ansatz. The gradient of ϕ is $x \mapsto (P \otimes I_p) x$ and, letting w_{ij} denote the weight of the edge (i, j) (with the convention that w_{ij} is zero whenever (i, j) is no edge), solving $u(x) = -(P \otimes I_p) x$ yields the so-called diffusive couplings

$$u_i(x) = \sum_{j=1}^{n} w_{ij}(x_j - x_i), \tag{2.25}$$

which admit implementation solely based on relative measurements. Moreover, achieving the scaling $u \mapsto ku$ can be realized by increasing the edge weights $w_{ij} \mapsto kw_{ij}$. We note that, for all $x \in \mathcal{D}$, $(f(x_1), \ldots, f(x_n))$ is itself a member of $\mathcal{D}$, which is the tangent space of $\mathcal{D}$ at x. However, as $\mathcal{D}$ is closed but unbounded, we must impose one of the aforementioned assumptions in order to proceed and guarantee existence of edge weights with the property that any larger edge weights will stabilize $\mathcal{D}$ asymptotically. Should the individual systems have asymptotically stable periodic orbits, then the n-fold Cartesian product of a closed subset of the region of asymptotic stability of these orbits with itself, call it U', is closed, invariant, and has compact intersection with $\mathcal{D}$, such that our framework can be applied to guarantee asymptotic stability of $\mathcal{D}$ relative to U' after noting that $\mathbb{R}^{np}$ is a tubular neighborhood of $\mathcal{D}$ (as $\mathbb{R}^{np}$ is a tubular neighborhood of any of its subspaces). Remarkably, we did not explicitly impose the constraint that u should rely on relative information only, but, however, arrived at a control that is structured according to (2.25). The connection between submanifold stabilization and structured controls will be strengthened in chapter 3.

Convergence and Performance

We did not yet discuss performance bounds and convergence rates of the closed loop. Although it is known that compact asymptotically stable invariant sets are uniformly attractive [9, Theorem 1.5.27], i.e. that for any given $\epsilon > 0$, there exists a $T > 0$ such that for all initial conditions in $V^{-1}([0,\alpha])$, for all $t > T$, $d(x(t),S) \leq \epsilon$, it is not always feasible to render T arbitrarily small. In the given setting, yet, it is possible to show that, for any $\epsilon > 0$ and any $T > 0$, there exists a $k_0 \geq 0$ such that for every $k > k_0$, for all initial conditions in $V^{-1}([0,\alpha])$, for all $t > T$, $d(x(t),S) \leq \epsilon$, as the following proposition, which was published in [M14], asserts.

Proposition 1. Let U be a tubular neighborhood of S. Let S be a critical submanifold of ϕ and let ϕ be regular on $U \setminus S$. If f is locally one-sided Lipschitz continuous and, for all $x' \in S$, the restriction of ϕ to the intersection of U with $F_{x'}$ is strongly convex, then, for every $\alpha > 0$ such that $V^{-1}(\{\alpha\})$ is contained in U, for all $\epsilon > 0$, for all $T > 0$, there exists a $k_0 \geq 0$ such that, for all $k > k_0$, all solutions $x : t \mapsto x(t)$ of (2.1) initialized in $V^{-1}([0,\alpha])$ satisfy $d(x(t),S) \leq \epsilon$ for all $t > T$.

Proof. Let k be greater than q/λ, with q and λ as in the previous proof. Pick any initial condition x_0 from $V^{-1}([0,\alpha])$ and reconsider the overestimate (2.15). Letting $x : t \mapsto x(t)$ denote the solution satisfying $x(0) = x_0$, compactness and invariance of $V^{-1}([0,\alpha])$ ensures that the domain of x is at least $[0,\infty)$ and application of the product rule yields that the derivative of $V \circ x$ is less than or equal to $2(q - k\lambda)(V \circ x)$. Applying Grönwall's inequality reveals that $V \circ x$ is less than or equal to

$$t \mapsto V(x_0) \, \mathrm{e}^{2(q-\lambda k)t} \, . \tag{2.26}$$

Recalling that $V(x_0) \leq \alpha$ and that V returns half the squared infimal distance of its argument to all points in S lets us solve this inequality for $t \mapsto d(x(t),S)$. Enforcing that the latter function should only attain values smaller than or equal to ϵ for all arguments exceeding T reveals that all k larger than

$$k_0 = \frac{1}{\lambda}\left(q - \frac{2\ln(\epsilon) - \ln(2\alpha)}{2T}\right) \tag{2.27}$$

provide this bound, establishing the assertion. $\qquad\square$

It is worth noting that, as a byproduct, the overestimate (2.26) also revealed that solutions of our controlled system approach the submanifold S at exponential rate.

The previous result does not only allow us to reach arbitrarily small neighborhoods of S in arbitrary time, i.e. tune the rate at which solutions approach S ad libitum, but also to tune other performance indices through k. For instance, let $J : \mathbb{R}^n \to \mathbb{R}$ be a nonnegative performance output, i.e. let the signal $t \mapsto J(x(t))$ be an indicator of the performance of the closed loop. Then, should J be continuous and zero at S, it is possible to attain the overestimate $J(x(T)) \leq \delta$ for any positive T and δ. Namely, as

$J^{-1}([0, \delta])$ is a neighborhood of S, there exists some $\epsilon > 0$ with the property that the implication

$$d\left(x\left(T\right), S\right) \leq \epsilon \quad \Rightarrow \quad J\left(x\left(T\right)\right) \leq \delta \qquad (2.28)$$

is true. Application of the foregoing theorem returns a k_0 such that all $k > k_0$ yield the desired performance. As such arbitrarily small performance bounds are possible for arbitrary continuous J vanishing on S, this observation suggests that controllers of the considered form are related to optimal control problems, and, more particular, raises the question whether our controllers with strongly convex integrals along the fibers F_x, $x \in S$, may be necessary for optimal submanifold stabilization, which we answer in the affirmative in chapter 3.

Example 3 (Navigation and Obstacle Avoidance). A relevant application in which one must indeed guarantee that some performance output can be underestimated is the navigation and obstacle avoidance problem for mechanical systems. A mechanical system in Euler-Lagrange form is described by the differential equation

$$\mathrm{M}\left(q\right) \ddot{q} + \mathrm{C}\left(q, \dot{q}\right) q = \mathrm{F}\left(q, \dot{q}\right) \qquad (2.29)$$

with q denoting the generalized positions, $\mathrm{M} : \mathbb{R}^n \to \mathbb{R}^{n \times n}$ being the inertia, $(q, \dot{q}) \mapsto \mathrm{C}(q, \dot{q}) q$ being the Coriolis / centrifugal forces, and $(q, \dot{q}) \mapsto \mathrm{F}(q, \dot{q})$ being the external forces and torques subject to our choice. For these systems, navigation and obstacle avoidance problems are of great interest. Navigation amounts to steering q towards a target submanifold S of $\mathbb{R}^n$, the destination, in a stable fashion and obstacle avoidance amounts to having empty intersection of $q([0, \infty))$ with certain subsets of $\mathbb{R}^n$, the obstacles. To solve both problems, one utilizes so-called navigation functions $\eta : \mathbb{R}^n \to \mathbb{R}$, which are zero at the target submanifold, maximal at the obstacles and regular everywhere else except for certain saddle points. These navigation functions can be constructed using the destination and obstacles as data [53]. It can then be shown that almost all solutions of the differential equation

$$\dot{q} = -k \nabla \eta\left(q\right), \qquad (2.30)$$

which are initialized away from the obstacles, approach the target submanifold in a stable fashion, whilst avoiding the obstacles. This behavior can also be enforced for (2.29) via the backstepping technique (cf. [83, section 6.1]), which, pretending that $\mathrm{C} : (q, \dot{q}) \mapsto 0$, returns the control

$$\mathrm{F}\left(q, \dot{q}\right) = -\mathrm{M}\left(q\right) \left(k\dot{q} + k \mathrm{H}_\eta\left(q\right) \dot{q} + \left(1 + k^2\right) \nabla \eta\left(q\right)\right), \qquad (2.31)$$

where $\mathrm{H}_\eta : \mathbb{R}^n \to \mathbb{R}^{n \times n}$ denotes the Hessian of η. Pretending that C vanishes is motivated by situations in which C is unknown or unmodeled, an issue which is usually resolved using adaptive controls (cf., e.g., [42]), thus involving implementation of certain differential equations and letting k vary with time. In contrast, we will show that it is sufficient to choose k sufficiently large but fixed. Viz., if one naively applies the aforementioned backstepping controller to the case where C is, in fact, nonzero, under the assumption that C is constant in its second argument, one arrives at the closed loop

$$\begin{bmatrix} \dot{x}_1 \\ \dot{x}_2 \end{bmatrix} = \begin{bmatrix} 0 \\ \mathrm{M}\left(x_1\right)^{-1} \mathrm{C}\left(x_1, x_2\right) x_1 \end{bmatrix} + \begin{bmatrix} -k & 1 \\ -1 & -k \end{bmatrix} \otimes I_n \begin{bmatrix} \nabla \eta\left(x_1\right) \\ x_2 \end{bmatrix} \qquad (2.32)$$

where $x_1 := q$ and $x_2 := \dot{q} + k\nabla\eta\,(q)$, i.e. the backstepping controller for the nominal (i.e. C being zero) system has turned the mechanical system (2.29) into a gradient system with drift, the drift vector field and scalar field being given by

$$f\,(x) = \begin{bmatrix} x_2 \\ \mathrm{M}\,(x_1)^{-1}\,\mathrm{C}\,(x_1,x_2)\,x_1 - \nabla\eta\,(x_1) \end{bmatrix}, \quad \phi\,(x) = \frac{1}{2}x_2 \cdot x_2 + \eta\,(x_1)\,, \qquad (2.33)$$

respectively, wherein ϕ is zero at $S \times \{0\}$ but positive elsewhere and the notation $x = (x_1, x_2)$ was employed. We remark that the skew-symmetric part of the matrix which is right-multiplied by $\nabla\phi$ in (2.32) reflects the Hamiltonian nature of the system whilst its symmetric part reflects the dissipation introduced by the backstepping controller. Assuming that η is strongly convex when restricted to $\{q\} + N_q S$, the restriction of ϕ to $\{x\} + N_x\,(S \times \{0\})$ is strongly convex, as well. As we outlined above, by increasing k sufficiently beyond some threshold k_0, the positions x_1 can not only be steered towards the target submanifold S in a stable fashion, but also can the function η be forced to attain arbitrarily small values in arbitrarily short time; this is due to the fact that $x \mapsto \eta\,(x_1)$ is continuous and vanishing on $S \times \{0\}$. In the present context, as η is maximal at the obstacles, we can guarantee obstacle avoidance despite the unknown C solely by tuning k and, in addition, guarantee that the positions x_1 are arbitrarily close to the destination or arbitrarily far away from the obstacles at arbitrary times. An implication of this observation is that one may always ignore C during backstepping design and merely tune k subsequently. A possible application of this approach is navigation and obstacle avoidance in unknown terrain.

In the present example, we applied the methods proposed in this section, in combination with the backstepping technique (for which we showed that it indeed brings the system to the form of a gradient system with drift), to solve navigation problems for mechanical systems. This approach is, in general, applicable to systems in strict feedback form. The example hence bridges the gap between the backstepping-based approach of El-Hawwary and Maggiore [35, section 5] to stabilization of closed sets and the approach pursued in this thesis.

The content of this example was previously published in [M11].

Asymptotic Stabilization of Compact Sets

In the above results, we assumed that S is a submanifold and we retain this property throughout the thesis. We only briefly mention that it is possible to consider compact invariant sets whose boundaries are submanifolds, as well. Namely, endow the boundary of the compact invariant set with the bundle structure we introduced for S and employ a Lyapunov function candidate which is equivalent to V for arguments not contained in the compact invariant set but zero elsewhere. The proof of Theorem 1 carries through except for the fact that S being an invariant set of the unforced system (2.3) implies that $f \circ r$ maps to tangent spaces of S, which in turn implies the equality $\nabla V \cdot (f \circ r) = 0$. This is to be replaced by Nagumo's theorem [67] which states that the compact set being an invariant set of the unforced system implies that $f \circ r$ maps to tangent cones of the invariant set, which in turn implies the inequality

$$\nabla V \cdot (f \circ r) \leq 0. \qquad (2.34)$$

This inequality is, however, also sufficient to complete the proof and thus does not obstruct us from considering compact invariant sets whose boundaries are themselves submanifolds.

Lipschitz Continuity

We assumed that f is locally one-sided Lipschitz continuous in Theorem 1, a condition which is sufficient to guarantee local forward uniqueness of solutions of the unforced system (2.3), cf. [11, section 1.11]. The more popular local Lipschitz continuity property employed in the Picard-Lindelöf theorem in order to guarantee local forward and backward uniqueness of solutions is a sufficient (but not necessary) condition for one-sided Lipschitz continuity and thus also suffices in order to prove Theorem 1. However, the definition of one-sided Lipschitz continuity requires an inner product between points and vectors as well as an additive operation between points, restricting its use to inner product spaces, whilst Lipschitz continuity merely requires a distance function, hence allowing its application in arbitrary metric spaces. In the next section, wherein we consider systems evolving on Riemannian manifolds (which are endowed with the properties of a metric space through application of the length functional to geodesic curves), we must hence restrict our attention to locally Lipschitz continuous f.

Invariance

The submanifold S was assumed to be an invariant set of the unforced system (2.3). Unless we allow for controls which are nonvanishing on S, and hence for possibly infinite energy of the signals $t \mapsto u_i(x(t))$, we have to retain this property as invariance is necessary for asymptotic stability. If we, however, do not insist to asymptotically stabilize S, but merely ask to asymptotically stabilize a set contained in an arbitrarily small neighborhood of S, then the invariance assumption becomes obsolete. Moreover, the conditions on ϕ relax significantly, as we claim in the following proposition, which was previously published in [M14].

Proposition 2. Let U be a neighborhood of S. Let S be a critical submanifold of ϕ, let ϕ be regular and positive on $U \setminus S$ and zero on S. If f is continuous, then, for every $\alpha > 0$ and $\epsilon > 0$ such that $\phi^{-1}([0, \alpha])$ is contained in U and $d(x, S) \leq \epsilon$ implies $\phi(x) < \alpha$, there exists a $k_0 \geq 0$ such that, for all $k > k_0$, the set of points for which $d(x, S) \leq \epsilon$ contains an asymptotically stable invariant set of (2.1) whose region of asymptotic stability contains $\phi^{-1}([0, \alpha])$.

Proof. As ϕ is continuous, zero on S, and positive elsewhere, there exists an $\delta > 0$ such that, for all x in the preimage of $[0, \delta]$ under ϕ, $d(x, S) \leq \epsilon$. Now choose such a δ which is also smaller than α. The Lie derivative of ϕ along $f - k\nabla\phi$ is given by the sum of the Lie derivatives $L_f\phi$ and $-kL_{\nabla\phi}\phi$ where the former equals $\nabla\phi \cdot f$ and the latter equals $-k\nabla\phi \cdot \nabla\phi$. Since f and $\nabla\phi$ are continuous, $x \mapsto \nabla\phi(x) \cdot f(x)$ assumes its maximum on the compact set $\phi^{-1}([\delta, \alpha])$, which we denote by q. As ϕ is regular on $U \setminus S$, $x \mapsto \nabla\phi(x) \cdot \nabla\phi(x)$ has a positive minimum on $\phi^{-1}([\delta, \alpha])$, which we denote by λ. Thus, for all k greater than $k_0 = q/\lambda$, the Lie derivative of ϕ along $f - k\nabla\phi$ is negative on the preimage of $[\delta, \alpha]$ under ϕ. This makes $\phi^{-1}([0, \delta])$ an invariant set and denoting the positive minimum of ϕ on $\phi^{-1}([\delta, \alpha])$ by c, we introduce the continuous function

$$x \mapsto \begin{cases} \phi(x) - c & \text{for } x \in U \setminus \phi^{-1}([0, \delta]) \\ 0 & \text{elsewhere} \end{cases} \tag{2.35}$$

which is zero on $\phi^{-1}([0, \delta])$, positive on $U \setminus \phi^{-1}([0, \delta])$, and has negative Lie derivative along $f - k\nabla\phi$ on $\phi^{-1}((\delta, \alpha])$, hence revealing that $\phi^{-1}([0, \delta])$ is an asymptotically stable invariant set through Lyapunov's direct method. Moreover, as $\phi^{-1}([0, \alpha])$ is compact and invariant, $\phi^{-1}([\delta, \alpha])$ is a subset of the region of asymptotic stability of $\phi^{-1}([0, \delta])$ by LaSalle's invariance principle. Recalling that the implication

$$\phi(x) \leq \delta \quad \Rightarrow \quad d(x, S) \leq \epsilon \tag{2.36}$$

holds true, this proves the claim. $\qquad\square$

Asymptotically stabilizing a set contained in an arbitrarily small neighborhood of S by choosing a parameter sufficiently large bares similarities with the definition of practical stability known from singularly perturbed systems (cf. [66]). Also, the fact that k_0 increases monotonically with decreasing ϵ relates the foregoing result to high-gain feedback (cf. [63]). In the light of this latter observation, in the case where S is a subspace of $\mathbb{R}^n$, our last proposition is similar in spirit to the almost controlled invariant subspaces problem [90, 91]. The proof resembles classical perturbation techniques [15].

Example 4 (Extremum Seeking with Drift). A possible application of the above result is to study extremum seeking with drift. Extremum seeking is a control technique with which one can find a control $u : \mathbb{R} \times \mathbb{R} \to \mathbb{R}^n$, solely requiring information about the current value of a function $\phi : \mathbb{R}^n \to \mathbb{R}$, which steers solutions of

$$\dot{x}(t) = u(\phi(x(t)), t) \tag{2.37}$$

as close to the minima of ϕ as one wishes, in a stable fashion (cf. [5]). In particular, one shows that an oscillatory control of the form

$$u(\phi(x), t) = \sum_{i=1}^{n} e_i \left(\phi(x) \sqrt{i\omega} \sin(i\omega t) - 2k\sqrt{i\omega} \cos(i\omega t) \right), \tag{2.38}$$

with $e_1, \ldots, e_n$ being the standard basis of $\mathbb{R}^n$ and $\omega > 0$ being a parameter with frequency interpretation, forces solutions of (2.37) into an $1/\omega$-neighborhood of the integral curves of $-k\nabla\phi$ in a stable fashion and hence allows one to bring solutions of (2.37) as close to the minima of ϕ as one wishes, by increasing ω. This can be proven using Lie bracket approximations (cf. [55]). If, however, the system (2.37) is corrupted by a drift vector field f, i.e. if one asks for a control u which steers solutions of

$$\dot{x}(t) = f(x(t)) + u(\phi(x(t)), t) \tag{2.39}$$

arbitrarily close to the minima of ϕ in a stable fashion, this technique is no longer applicable. By similar techniques as the ones mentioned above (viz. Lie bracket approximation), one can however show that (2.38) lets solutions of (2.39) remain in an $1/\omega$-neighborhood of the integral curves of $f - k\nabla\phi$. By the foregoing proposition, we moreover know that for every arbitrarily small neighborhood of the minima of ϕ, there exists a k such that integral curves of $f - k\nabla\phi$ approach the given neighborhood in a stable fashion. Combining these two results reveals that, for every arbitrarily small ϵ-neighborhood of the minima of ϕ, the tuple (k, ω) can be chosen such that solutions of (2.39) approach this neighborhood under the control (2.38) in a stable fashion. This is precisely done by first choosing a subneighborhood of the ϵ-neighborhood in which one forces integral curves of $f - k\nabla\phi$ by tuning k to subsequently increase ω enough such that a neighborhood of the subneighborhood, contained in the target ϵ-neighborhood, attracts solutions of (2.39) under the extremum seeking feedback (2.38).

The content of this example was previously published in [M16].

Example 5 (Practical Synchronization). Proposition 2 can, in contrast to the approach taken in Example 2, also be utilized to investigate synchronization problems for non-identical systems, i.e.

$$\dot{x}_i = f_i(x_i) + u_i(x), \quad i = 1, \ldots, n \tag{2.40}$$

even if the systems should have no mutual internal model and (exact) synchronization is thus impossible (cf. [89]). Namely, in the language employed before, the diagonal $\mathscr{D}$ is no invariant set of the unforced system in this case, as there exists $x \in \mathscr{D}$ for which $(f_1(x_1), \ldots, f_n(x_n))$ is not a member of $\mathscr{D}$ (which is itself the tangent space of $\mathscr{D}$ at x). In this case, however, by the foregoing proposition, if the vector fields $f_1, \ldots, f_n$ are continuous, for every $\epsilon > 0$, our weights can be scaled, i.e. $w_{ij} \mapsto kw_{ij}$, such that the diffusive couplings (2.25) render an invariant subset of an ϵ-neighborhood of $\mathscr{D}$ asymptotically stable relative to some closed invariant set having compact intersection with $\mathscr{D}$. This convergence behavior has been coined practical synchronization and was successfully applied to robust synchronization and cluster synchronization problems.

Remarkably, Pecora and Carroll [70] experimentally observed that nonidentical dif-fusively coupled chaotic circuits have small ultimate bounds on the differences of their evolutions, a phenomenon that can be interpreted with knowledge about the present example.

The content of this example was previously published in [M19, M10, M12].

Computational Approximation of the Constructed Control

Our results are implicit in the sense that they prove existence of a parameter k_0 having the properties mentioned in Theorem 1. Though k_0 was then explicitly constructed in the proof, we employed the maxima and minima of certain parameters (q_x and λ_x, respectively) to do so. Explicitly solving these minimizations and maximizations in practice could turn out to be a hard task. It is thus remarkable that k_0 can be approximated with arbitrary precision via pointwise evaluation of the ratio between the Lie derivatives $L_f V$ and $L_{\nabla\phi} V$, viz. a necessary and sufficient condition for k to render the Lie derivative (2.10) negative away from S is that it exceeds the smallest k_0 sufficing

$$k_0 \geq \frac{f(x) \cdot \pi(x)}{\nabla\phi(x) \cdot \pi(x)} \qquad \forall x \in U \setminus S. \tag{2.41}$$

The underestimate for k_0 stemming from the ratio of these Lie derivatives allows to be evaluated on each point of some finite ϵ-net of the compact set $V^{-1}([0, \alpha])$ (for instance obtained from farthest-first traversal or Lloyd's algorithm / Voronoi iteration) and thus can be approximated algorithmically with arbitrary precision (the precision increases as ϵ decreases), viz. by subsequently assigning the above underestimate for k_0 at x to k_0 for every point on the ϵ-net. Furthermore, this quantity can turn out significantly smaller than the quantity q/λ suggested in the proof, for instance in cases where $f(x)$ is in $T_{r(x)}S$ everywhere and k_0 could hence be set to zero.

Example 6 (Formation Control). Formation control problems can be solved via the methods from this section and, in particular, a drift vector field f for which the afore-mentioned condition $f(x) \in T_{r(x)}S$ holds will arise, letting $k_0 = 0$. In formation control problems, one wants a group of systems to eventually attain a configuration of given shape (cf., e.g. [25]). Letting the shape and position of the desired configuration be determined by a submanifold S of $\mathbb{R}^p$, formation control amounts to bringing the systems to an equidistant configuration on S. Assuming that S is Riemannian, let $d : S \times S \to \mathbb{R}$ be the distance function on S. Consider a formation control problem for n systems, denote $x = (x_1, \ldots, x_n)$, choose an undirected, weighted graph with n vertices and w_{ij} denoting the weight of edge (i, j) (with the convention that w_{ij} is zero if (i, j) is no edge) and consider the function

$$\theta : S^n \to \mathbb{R}, \quad x \mapsto \sum_{i=1}^{n} \sum_{j=1}^{n} \frac{w_{ij}}{4} d(x_i, x_j)^2 \tag{2.42}$$

whose maximizers are often referred to as balanced configurations (cf. [81]), and which contain the equidistant configurations of x on S for appropriate graphs. Further, let U be a tubular neighborhood of S and define

$$\phi : U^n \to \mathbb{R}, \quad x \mapsto \sum_{i=1}^{n} \|\pi(x_i)\|^2 \tag{2.43}$$

whose minimizers are just the members of S^n. Let $\operatorname{grad} \theta$ denote the gradient vector field of θ, taking each point $x \in S^n$ to a vector in $T_x S^n$. Let $r : U^n \to S^n$ denote the retraction onto S^n. Then the solutions of

$$\dot{x} = \operatorname{grad} \theta(r(x)) - k\nabla\phi(x) \tag{2.44}$$

approach a balanced configuration on S in a stable fashion for any k greater than $k_0 = 0$, thus solving the formation control problem. We remark that the latter is a gradient system with drift, with the drift being the composition of the gradient of θ with the retraction onto S^n. Denoting the drift by $f = (\operatorname{grad} \theta) \circ r$, we have the condition $f(x) \in T_{r(x)}S^n$ satisfied, explaining why k_0 is zero in this example. The convergence properties of the differential equation (2.44) are depicted in Figure 2.5, wherein its solutions for some initial condition, labeled by blue circles, are plotted in blue. As for the underlying graph, we chose the cycle graph with unit weights. On the left-hand side of the figure, S is $\mathbb{S}_1$, the unit circle, embedded in $\mathbb{R}^2$. On the right-hand side, S is the unit triangle, the polygon resulting from rotating a unit vector twice by $2\pi/3$. For the unit triangle, the corners were locally smoothed out to make it a differentiable manifold. In both figures, the submanifolds are plotted in black. The limiting points

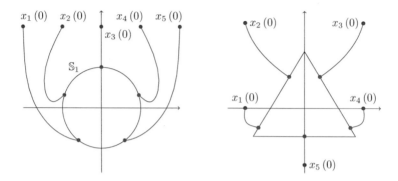

Figure 2.5: Convergence to a circular (left) and a triangular (right) formation

of the solutions are indicated by red circles. We observe that the depicted solutions eventually attain an equidistant configuration on the target submanifold, which was the goal in the formation control problem we posed.

The content of this example was previously published in [M20].

Summary

In this section, we presented a method for rendering an invariant submanifold of an input affine system asymptotically stable. Our approach was based upon choosing controls such that the closed loop becomes a gradient system with drift. This allowed us to cast the sufficient conditions for asymptotic stabilizability in terms of the scalar field defining the gradient vector field, namely requiring strong convexity of its restriction to the affine spaces resulting from translating the normal spaces of our submanifold. The controller design problem is thereby shifted to designing a scalar field, which can be achieved in a constructive fashion. The applicability of our result is inherently restricted to tubular neighborhoods of the given submanifold. We discussed how the assumptions that $g_1 u_1 + \cdots + g_m u_m$ is integrable, that S is compact, that S is a submanifold, and that S is an invariant set of the unforced system can be relaxed as well as how local one-sided Lipschitz continuity may be replaced by local Lipschitz continuity. Further, we elaborated how certain convergence rates / performance bounds can be achieved and how the parameter k_0 can be approximated algorithmically. We presented different applications of our approach.

Discussion

If one defines π as an output of the system (2.1), then the submanifold stabilization problem can be cast as an output stabilization problem, for the value of π vanishes if and only if its argument is contained in S. Output stabilization for linear systems has been studied in detail by Wonham [94, Chapter 4] (or [10]). The more general output regulation problem was solved by Francis [41] for linear systems and by Isidori and

Byrnes [49] for nonlinear systems. The output regulation machinery can be applied to submanifold stabilization problems by defining an exosystem whose solution evolves on S and letting the regulation error be the deviation of the state of our system from the state of the exosystem; a possible shortcoming of this approach is that the exosystem is usually required to have Poisson stable initial conditions. If one thus was to implement the solution to the submanifold stabilization problem based upon output regulation on a device, then small perturbations will drive the exosystem, and hence the system, away from the desired submanifold for all future times.

It shall be emphasized that the pattern generation problem, as it was formulated by Brockett [20], as well as the path following problem, as formulated by Nielsen and Maggiore [68], both fit well into the framework proposed in this section.

2.2 Asymptotic Stabilization of Submanifolds of Riemannian Manifolds

In the foregoing section, we developed methods for submanifold stabilization in input affine systems (2.1) whose solutions evolve in $\mathbb{R}^n$. These methods are not yet applicable to systems whose state spaces are different from $\mathbb{R}^n$, for instance to rigid bodies, which evolve on the special Euclidean group, rotations, which live on the special orthogonal group, or oscillators, which move on homotopy circles. A commonality of these state spaces is that they are all Riemannian manifolds. In this section, we thus study the problem of asymptotically stabilizing a submanifold S of some Riemannian manifold M which is invariant under (2.1), i.e. $x \mapsto (x, f(x))$ and $x \mapsto (x, g_1(x)), \ldots, x \mapsto (x, g_m(x))$ are smooth sections of TM, the tangent bundle of M. We again assume that S is compact, smoothly embedded, and an invariant set of the unforced system (2.3), i.e., $x \mapsto (x, f|_S(x))$ is a smooth section of TS, the tangent bundle of S (a subbundle of TM). The integrability Ansatz here reads

$$\sum_{i=1}^m g_i u_i = -k \operatorname{grad} \phi, \tag{2.45}$$

for some continuously differentiable function $\phi : M \to \mathbb{R}$ with gradient vector field $\operatorname{grad} \phi$, i.e. $x \mapsto (x, \operatorname{grad} \phi(x))$ is a smooth section of TM. The main results of this section were published in [M3].

Letting $T_x S$ denote the tangent space of S at x, the normal space of S at x, denoted by $N_x S$, is the orthogonal complement of $T_x S$ in $T_x M$ (the tangent space of M at x) with respect to the Riemannian metric. The normal bundle NS of S, defined by (2.5), is a vector bundle over S composed of the fibers $N_x S$. Yet, in order to proceed as in the foregoing section and define a meaningful notion of convexity, we must also define an intrinsic fiber structure on M (i.e. one whose fibers are submanifolds of M), which we next introduce using the exponential map: let $\exp : TM \to M$ denote the exponential map, then a tubular neighborhood of S is the diffeomorphic image of the restriction of $\exp$ to NS. We ask for the restriction of $\phi : M \to \mathbb{R}$ to the intersection of a tubular neighborhood with the images of $(x, N_x S)$ under $\exp$, $F_x := \exp(x, N_x S)$, to be strongly geodesically convex (with strongly geodesically convex, we mean that $\phi \circ \gamma$ is strongly convex in the usual sense for any unit speed, i.e. arc length parametrized, geodesic γ). Let U be a tubular neighborhood of S. With $\mathsf{P}_1 : NS \to S$, $(x, v) \mapsto x$ being the bundle projection and $\mathsf{P}_2 : (x, v) \mapsto v$ being the fiber projection,

$$r := \mathsf{P}_1 \circ \left(\exp\big|_{NS}\right)^{-1} : U \to S \tag{2.46}$$

is a smooth retraction onto S and $\pi := \mathsf{P}_2 \circ \left(\exp\big|_{NS}\right)^{-1}$ is its counterpart, returning a vector in $N_{r(x)} S$ for a given point x in U, and completing the identity

$$\exp(r, \pi) = \operatorname{id}, \tag{2.47}$$

where $\operatorname{id} : U \to U$, $x \mapsto x$. In other words, the restriction of the exponential map to the normal bundle of S takes the role of the map μ from the previous section, which we

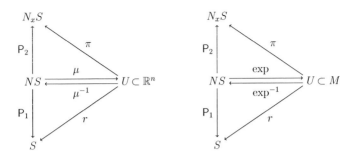

Figure 2.6: Commutative diagrams for section 2.1 (left) and section 2.2 (right)

illustrated in Figure 2.6, wherein the commutative diagram on the left, which was valid in section 2.1, is the same as the commutative diagram on the right, which is valid in the present section, except for μ having been replaced by exp (strictly speaking, $N_x S$ would have to be replaced by the union of all $N_x S$ over all x in S in the commutative diagrams).

As the magnitude of the value of the function π grows with the distance of its argument to S, it defines the "natural" Lyapunov function candidate

$$V := \frac{1}{2} \langle \pi, \pi \rangle : U \to \mathbb{R}, \tag{2.48}$$

which is smooth, vanishing on S, and positive elsewhere. Here, $\langle \cdot, \cdot \rangle$ denotes the Riemannian metric. Assuming (2.45), the following theorem, which was previously published in [M3], is the main result of this section.

Theorem 2. Let U be a tubular neighborhood of S. Let S be a critical submanifold of ϕ and let ϕ be regular on $U \setminus S$. If f is locally Lipschitz continuous and, for all $x \in S$, the restriction of ϕ to the intersection of U with F_x is strongly geodesically convex, then, for every $\alpha > 0$ such that $V^{-1}(\{\alpha\})$ is contained in U, there exists a $k_0 \geq 0$ such that, for all $k > k_0$, S is an asymptotically stable invariant set of (2.1) and $V^{-1}([0, \alpha])$ is a subset of its region of asymptotic stability.

Proof. The Lie derivative of V along $f - k \operatorname{grad} \phi$ is given by

$$L_{f - k \operatorname{grad} \phi} V = L_f V - k L_{\operatorname{grad} \phi} V, \tag{2.49}$$

hence allowing us to investigate the Lie derivatives $L_f V$ and $L_{\operatorname{grad} \phi} V$ separately.

First, consider the Lie derivative of V along f. As $x \mapsto (x, f|_S)$ is a smooth section of TS, r maps to S, and π maps to the fibers of NS, we have

$$\langle \pi, f \circ r \rangle = 0. \tag{2.50}$$

Knowing that $\operatorname{grad} V = \Gamma_r^{\mathrm{id}} \pi$, wherein $\Gamma_x^{x'} : T_x M \to T_{x'} M$ denotes the parallel transport map, we thus conclude

$$L_f V = \langle \Gamma_r^{\mathrm{id}} \pi, f - \Gamma_r^{\mathrm{id}} f \circ r \rangle \leq V + \frac{1}{2} \langle f - \Gamma_r^{\mathrm{id}} f \circ r, f - \Gamma_r^{\mathrm{id}} f \circ r \rangle \tag{2.51}$$

with the equality following from (2.50) by parallel transport and the underestimate following from Young's inequality. As f is locally Lipschitz continuous, for every x, there exists an open neighborhood U_x of x and a constant q_x such that, for all x' in U_x, the overestimate

$$\langle f(x) - \Gamma^x_{x'} f(x'), f(x) - \Gamma^x_{x'} f(x') \rangle^{1/2} \leq q_x d(x, x') \tag{2.52}$$

holds true, wherein, $d : M \times M \to \mathbb{R}$ is the distance function obtained from application of the length functional to geodesic curves. Now choose some $\alpha > 0$ such that $V^{-1}(\{\alpha\})$ is contained in U. As S is compact, $V^{-1}([0, \alpha])$ is compact and hence its open cover $\{U_x\}_{x \in U}$ has finite subcover, over whom a maximal q_x exists, which we denote by q. Recalling that $\langle \pi, \pi \rangle = d(\mathrm{id}, r)^2$ leaves us with the overestimate

$$L_f V \leq (1 + q) V \tag{2.53}$$

for the Lie derivative of V along f.

Next, we turn our attention to the Lie derivative of V along $\mathrm{grad}\,\phi$. We had assumed that the restriction of ϕ to the intersection of U with F_x is strongly geodesically convex. Thus, following from the chain rule, for every x in S, there exists a $\lambda_x > 0$, varying continuously with x, such that for all $x' \in F_x$, for all $s \geq s'$,

$$\left\langle \left((\mathrm{grad}\,\phi) \circ \gamma^{x'}_x\right)(s), \dot{\gamma}^{x'}_x(s) \right\rangle - \left\langle \left((\mathrm{grad}\,\phi) \circ \gamma^{x'}_x\right)(s'), \dot{\gamma}^{x'}_x(s') \right\rangle \geq \lambda_x(s - s'), \tag{2.54}$$

where $\gamma^{x'}_x : [0, d(x, x')] \to M$ denotes the unit speed geodesic (i.e. with arc length parametrization) joining $x = \gamma^{x'}_x(0)$ and $x' = \gamma^{x'}_x(d(x, x'))$. As the assignment $x \mapsto \lambda_x$ is continuous and S is compact, some minimal λ_x exists, which we denote by λ. We note that, for a given x in U, the image of the unit speed geodesic $\gamma^{r(x)}_x$ is contained in $F_{r(x)}$. Thus, the underestimate

$$-\langle \mathrm{grad}\,\phi, \dot{\gamma}^r_{\mathrm{id}}(0) \rangle \geq \lambda d(\mathrm{id}, r), \tag{2.55}$$

is valid on U, where we exploited that S is a critical submanifold of ϕ, and that hence $(\mathrm{grad}\,\phi) \circ r$ sends every point in U to the zero vector. As for the Lie derivative of V along $\mathrm{grad}\,\phi$, this leaves us with the underestimate

$$L_{\mathrm{grad}\,\phi} V \geq 2\lambda V, \tag{2.56}$$

following from application of $\mathrm{grad}\,V = \Gamma^{\mathrm{id}}_r \pi$ and $\pi = d(\mathrm{id}, r)\,\dot{\gamma}^{\mathrm{id}}_r(0)$.

Combining the overestimate (2.53) for the Lie derivative of V along f and the underestimate (2.56) for the Lie derivative of V along $\mathrm{grad}\,\phi$ with (2.49) yields the overestimate

$$L_{f - k\,\mathrm{grad}\,\phi} V \leq (1 + q - 2k\lambda) V \tag{2.57}$$

for the Lie derivative of V along $f - k\,\mathrm{grad}\,\phi$. Letting k_0 be $(1 + q)/2\lambda$, we figure that, for any k greater than k_0, the Lie derivative of V along $f - k\,\mathrm{grad}\,\phi$ becomes negative for all x in $V^{-1}([0, \alpha])$ which are not in S. As $V^{-1}([0, \alpha])$ is a neighborhood of S and S is compact, this reveals that S is an asymptotically stable invariant set by Lyapunov's direct method. We thus also know that this Lie derivative is nonpositive on the compact set $V^{-1}([0, \alpha])$, making it forward invariant. Hence, by LaSalle's invariance principle, $V^{-1}([0, \alpha])$ is a subset of the region of asymptotic stability of S, proving the assertion. $\qquad\square$

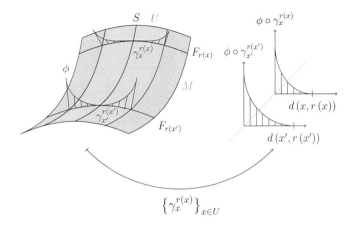

Figure 2.7: The geodesics $\gamma_x^{r(x)}$ bring ϕ to a family of strongly convex functions

The key idea of the proof is to work with the fibers $F_x \cap U$ on which we asked ϕ to be strongly geodesically convex. The members of the family of geodesics $\gamma_x^{r(x)}$, $x \in U$, have images, all of whom are contained in the aforementioned fibers. It is thus fair to say that the family of geodesics $\gamma_x^{r(x)}$, $x \in U$, brings the scalar field ϕ to a family of strongly convex functions, which we illustrated in Figure 2.7. On the left-hand side, we see two restrictions of the scalar field ϕ, plotted in blue, to fibers of the form $F_x \cap U$. The according submanifolds F_x are depicted in black, and so is S itself. Within the fibers, two geodesics $\gamma_x^{r(x)}$ are plotted in red, with their endpoints represented by red circles. We highlighted a portion of the tubular neighborhood U in blue and a portion of the ambient manifold M in red. On the right-hand side, we plotted the composition of ϕ with the two geodesics, in blue. The dotted line suggests that we may attach such a strongly convex graph to any point in S.

Again, the constructive implication of this graphical interpretation of our result is that one merely needs to design strongly geodesically convex functions on the fibers F_x. For instance, a finite ϵ-net of S could be chosen and strongly geodesically convex functions on each fiber intersecting S at a point of the ϵ-net could be designed, between whom one could subsequently interpolate smoothly. Should the codimension of S in M, i.e. the dimension of the fibers F_x, be 1, then this can be done by constructing strongly convex endomorphisms of $\mathbb{R}$ and pulling them back to M via the geodesics $\gamma_x^{r(x)}$. Should the codimension of S in M, and thus the dimension of the fibers F_x, be larger than 1, then one would require charts for the submanifolds F_x. Having such charts at hand, strongly geodesically convex functions on the fibers could be constructed via first constructing strongly convex functions whose dimension is the codimension of S in M and consequently pulling these functions back to M via the inverse of the charts. Again, it suffices to construct such a strongly geodesically convex restriction of ϕ to one

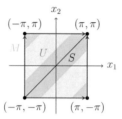

Figure 2.8: State space and convergence properties of two Kuramoto oscillators

fiber, say F_x, to then arrive at another restriction, say to $F_{x'}$, via

$$\phi\big|_{F_{x'}} : x'' \mapsto (c \circ r)(x'')\left(\phi\big|_{F_x} \circ \exp\right)(x, \operatorname{rot} \pi(x'')), \qquad (2.58)$$

wherein c is some scalar field on S which is uniformly bounded from below by a positive scalar and rot is a linear isometry such that $\operatorname{rot} N_{x'}S = N_xS$.

Remarkably, the lower bound for k constructed in the proof is $k_0 = (1 + q)/2\lambda$ whereas we arrived at $k_0 = q/\lambda$ in the foregoing section. The distinction between the two theorems was that Theorem 1 assumed f to be one-sided Lipschitz continuous while we assumed Lipschitz continuity in Theorem 2. Thus, the parameter q from the last section could, potentially, be negative (which is indeed the case for contracting vector fields) whilst the parameter q appearing in this section is nonnegative. As a consequence, k_0 is here larger than or equal to $1/2\lambda$ and can hence not be zero by construction. However, although the above, conservative, construction of k_0 yields a positive value, k_0 may be zero under the condition that the Lie derivative of V along f is zero, i.e. if $\langle \Gamma_r^{\mathrm{id}}\pi, f \rangle$ is zero.

Example 7 (The Kuramoto Model). In studying oscillator synchronization, the Kuramoto model [54] has become a canonical example. Therein, the so-called Kuramoto oscillators, with frequency ω, are the differential equations

$$\dot{x}_i = \omega + k \sum_{j=1}^{n} \sin(x_j - x_i), \quad i = 1, \ldots, n \qquad (2.59)$$

with state spaces $[-\pi, \pi)$, where π is identified with $-\pi$. For $n = 2$, the resulting product manifold M is depicted in Figure 2.8 as its fundamental polygon (which is the fundamental polygon of the torus), with the vertices plotted as black circles and the edges plotted as colored arrows, the mutual association of edges indicated by the color of the arrows. Proceeding as in [84], the Kuramoto oscillators can be brought to the form of a gradient system with drift, where drift vector field and scalar field are given by

$$f(x) = \omega 1_n, \quad \phi(x) = \frac{n^2}{2} - \frac{1}{2}\sum_{i=1}^{n}\sum_{j=1}^{n}\cos(x_i - x_j), \qquad (2.60)$$

respectively, wherein 1_n is the n-fold vector of ones. We recall that synchronization is characterized by convergence to the diagonal; the drift vector field f leaves the diagonal

$\mathcal{D}$ of the product space M, depicted in Figure 2.8 as a black line, invariant and the function ϕ is zero on $\mathcal{D}$ but positive elsewhere. Furthermore, ϕ is strongly convex when restricted to the normal spaces of $\mathcal{D}$ within a tubular neighborhood U of $\mathcal{D}$. This is verified by analyzing its Hessian

$$
\begin{bmatrix}
\sum\limits_{i=2}^{n} \cos{(x_i - x_1)} & -\cos{(x_2 - x_1)} & \cdots & -\cos{(x_n - x_1)} \\
-\cos{(x_1 - x_2)} & \sum\limits_{\substack{i=1 \\ i \neq 2}}^{n} \cos{(x_i - x_2)} & & \\
\vdots & & \ddots & \\
-\cos{(x_1 - x_2)} & & & \sum\limits_{i=1}^{n-1} \cos{(x_i - x_n)}
\end{bmatrix}
\tag{2.61}
$$

whose nullspace is $\mathcal{D}$ and whose eigenvalues (except the zero eigenvalue) are uniformly bounded from below by positive scalars in a neighborhood U of $\mathcal{D}$, implying that the restriction of ϕ to normal spaces of the diagonal $\mathcal{D}$ is indeed strongly convex. As $L_f V$ is zero everywhere, we thus know that $\mathcal{D}$ becomes asymptotically stable for any k greater than $k_0 = 0$. For $n = 2$, the neighborhood U of the diagonal on which the restriction of ϕ to normal spaces of S is strongly convex is characterized by the neighborhood on which the nonzero eigenvalue of the Hessian of ϕ is uniformly bounded from below by a positive scalar. As this eigenvalue is given by $2 \cos{(x_1 - x_2)}$, we conclude that U is given by

$$
U = \{(x_1, x_2) \,|\, |x_1 - x_2| < \pi/2\},
\tag{2.62}
$$

which is highlighted blue in Figure 2.8. Although this does not characterize the convergence properties of the Kuramoto oscillators completely (for the latter, cf. [84]), application of our framework however significantly simplified the analysis of the stability properties of $\mathcal{D}$.

The content of this example was previously published in [M3].

The applicability of Theorem 2 is inherently limited to tubular neighborhoods of S, i.e., the underestimate provided for its region of asymptotic stability in the theorem must be contained in one of its tubular neighborhoods. We omitted the discussion on how large these sets may be in the foregoing section as we are only now (considering Riemannian manifolds) able to connect this discussion to topological properties of M. In the foregoing section, where $M = \mathbb{R}^n$, we could find arbitrarily large tubular neighborhoods, for instance for submanifolds S being subspaces of $\mathbb{R}^n$. For S being a homotopy circle, tubular neighborhoods had to at least exclude one point located inside S, in consistency with the Poincaré-Bendixson theorem. Now that we admitted for state spaces which are Riemannian manifolds, the situation becomes more delicate: requiring injectivity of the restriction of the exponential map to the normal bundle of S, we find that compact M will usually contain points which we cannot trace back uniquely to S along the fibers F_x. The set of such points can be figured from the intersection of the cut loci of x with the fiber F_x, i.e. the number of cut loci will provide an estimate of how many points have to be excluded from U. Also, Hopf index theory and the Poincaré-Hopf theorem will not allow for an arbitrary number of equilibria, depending on the Euler characteristic of M. More generally, Conley indices may predict existence of isolated invariant sets away from S. Another topological restriction stems from the requirement that ϕ was asked to be regular on U without S. In particular, the Euler characteristic of M and its Betti numbers provide bounds for how many critical points ϕ must have in connection with Morse / Morse-Bott theory. Further, the Lusternik-Schnirelmann category of M provides information about how many critical points ϕ must at least have, and thus, in turn, how many points we have to exclude from U in order to satisfy the regularity assumptions of the theorem. However, in particular cases, U can be chosen to only exclude a set of measure zero (when measured with the Riemannian measure). One example for manifolds for which this is, in general, possible, are Blaschke manifolds, as their elements have equidistant cut loci, the distance all of whom to the point is precisely the diameter of M.

Example 8 (Almost Global Oscillator Synchronization). Characterizing the region of asymptotic stability of the diagonal $\mathscr{D}$ for the Kuramoto oscillators is a hard task (cf. [84]). Yet, using the methods from this section, it turns out comparatively simple to construct coupled oscillators, living on the circle $\mathbb{S}_1$ (embedded in $\mathbb{R}^2$), whose solutions synchronize for almost all initial conditions. In particular, as the circle is a Blaschke manifold, every neighborhood of a point on the circle that excludes its cut locus, which is merely a single (opposite) point, is tubular and free from critical points of the squared distance function, except for the point itself (cf. [31]). Thus, the diagonal of the n-fold Cartesian product of the circle with itself, the n-torus $M = \mathbb{S}_1^n$, has a tubular neighborhood U with the property that $\mathbb{S}_1^n \setminus U$ has measure zero (when measured with the Riemannian measure). Furthermore, $U \setminus \mathscr{D}$ is free of critical points of

$$\phi : x \mapsto \frac{1}{2} d\left(x, r\left(x\right)\right)^2 = \frac{1}{4} \sum_{i=1}^{n} \sum_{j=1}^{n} \mathrm{acos}\left(x_i \cdot x_j\right)^2 \tag{2.63}$$

wherein r is the retraction from U onto $\mathscr{D}$ and the notation $x = (x_1, \ldots, x_n)$ was employed. In order for the oscillators to move on the circle with frequency ω, the drift vector field of an individual oscillator should be $x_i \mapsto \omega \Omega x_i$, with Ω the infinitesimal generator of $\mathfrak{so}\left(2\right)$, the Lie algebra of $\mathrm{SO}\left(2\right)$. The resulting gradient system with drift is hence given by

$$\dot{x}_i = \omega \Omega x_i + k \sum_{j=1}^{n} \mathrm{acos}\left(x_i \cdot x_j\right) \log\left(\begin{bmatrix} x_i & \Omega x_i \end{bmatrix}^\top \begin{bmatrix} x_j & \Omega x_j \end{bmatrix}\right) x_i \tag{2.64}$$

where $\log : \mathrm{SO}\left(2\right) \to \mathfrak{so}\left(2\right)$ is the logarithmic map from the special orthogonal group $\mathrm{SO}\left(2\right)$ to its Lie algebra $\mathfrak{so}\left(2\right)$. Not only does the diagonal $\mathscr{D}$ become an asymptotically stable invariant set of (2.64) for any k greater than $k_0 = 0$, but, moreover, any sublevel set of V contained in a tubular neighborhood U of $\mathscr{D}$ with the property that ϕ is regular on $U \setminus \mathscr{D}$ is a subset of the region of asymptotic stability of $\mathscr{D}$. By the above reasoning, based on the fact that $\mathbb{S}_1$ is a Blaschke manifold, there exists a tubular neighborhood U with $\mathbb{S}_1^n \setminus U$ having measure zero. Further, every sublevel set of V, except for the one for which V is half the squared diameter of M, is contained in U. Hence, solutions of (2.64) approach $\mathscr{D}$ asymptotically in a stable fashion for almost all initial conditions. The content of this example was previously published in [M3].

Example 9 (Almost Global Attitude Synchronization). The technique from the foregoing example can be applied to attitude synchronization of rigid bodies (cf. [79]), which is relevant, e.g., in satellite formations. More particular, as the special orthogonal group $SO(3)$ is diffeomorphic to the real projective space $\mathbb{RP}^3$, which itself is a Blaschke manifold,

$$\phi : x \mapsto \frac{1}{2} d\left(x, r\left(x\right)\right)^2 = -\frac{1}{8} \sum_{i=1}^{n} \sum_{j=1}^{n} \text{tr}\left(\log\left(x_i^\top x_j\right)^2\right) \tag{2.65}$$

where $\log : SO(3) \rightarrow \mathfrak{so}(3)$ is the logarithmic map from the special orthogonal group $SO(3)$ to its Lie algebra $\mathfrak{so}(3)$, is regular on $U \setminus \mathcal{D}$, where $\mathcal{D}$ is the diagonal of $SO(3)^n$ and U is a tubular neighborhood of $\mathcal{D}$ with the property that $SO(3)^n \setminus U$ has measure zero in $SO(3)^n$ (measured with the Riemannian measure). In the previous example, the drift was such that $L_f V = 0$ everywhere, letting k_0 be zero. In the present example, consider n rotations, evolving on $SO(3)^n$, each of whom is subject to the drift

$$f\left(x_i\right) = \left(1 + \sqrt{-\frac{1}{2} \text{tr}\left(\log\left(x_i\right)^2\right)}\right) x_i \left(\omega_1 \Omega_1 + \omega_2 \Omega_2 + \omega_3 \Omega_3\right) \tag{2.66}$$

where ω_1, ω_2, ω_3 are scalars and Ω_1, Ω_2, Ω_3 are the infinitesimal generators

$$\Omega_1 = \begin{bmatrix} 0 & 0 & 0 \\ 0 & 0 & 1 \\ 0 & -1 & 0 \end{bmatrix}, \; \Omega_2 = \begin{bmatrix} 0 & 0 & -1 \\ 0 & 0 & 0 \\ 1 & 0 & 0 \end{bmatrix}, \; \Omega_3 = \begin{bmatrix} 0 & 1 & 0 \\ -1 & 0 & 0 \\ 0 & 0 & 0 \end{bmatrix} \tag{2.67}$$

of the Lie algebra $\mathfrak{so}(3)$. The magnitude of the drift vector field scales with the distance of an individual rotation to the identity element of $SO(3)$. The integral curves of f contained in $\mathcal{D}$ are nontrivial, i.e. $\mathcal{D}$ is no submanifold of equilibria. Here, we find that the diagonal $\mathcal{D}$ becomes an asymptotically stable invariant set, i.e. attitude synchronization is achieved, for k greater than

$$k_0 = 2\sqrt{\omega_1^2 + \omega_2^2 + \omega_3^2}. \tag{2.68}$$

Indeed, in Figure 2.9, two solutions of the gradient system with drift (associated with the scalar field (2.65) and the drift vector fields (2.66)) are depicted for $n = 2$ rotation matrices, multiplied from the right with a box in $\mathbb{R}^3$ in order to visualize the rotation. One of the boxes is plotted in blue whilst the other box is plotted in red. The parameters ω_1, ω_2, ω_3 were set to $\omega_1 = 3$, $\omega_2 = 2$, $\omega_3 = 1$. In the top row, k was chosen positive but smaller than k_0 and it can be seen that the two rigid bodies eventually do not synchronize (the rigid bodies are illustrated at times ranging from 0 (far left) to 50 (far right)). In the bottom row, k was chosen greater than k_0 and it can be seen that the two rigid bodies eventually synchronize (the rigid bodies are illustrated at times ranging from 0 (far left) to 10 (far right)). In both simulations, the same initial conditions were chosen.

The content of this example was previously published in [M13].

$k < k_0$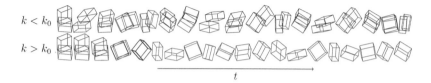
$k > k_0$

$$t$$

Figure 2.9: Synchronization of rigid bodies

An Algebraic Characterization of Asymptotic Stabilizability

So far, we assumed that $g_1 u_1 + \cdots + g_m u_m$ is integrable throughout the chapter. In particular, we assumed the integrability conditions (2.4) and (2.45) in the previous and present sections, respectively. In the previous section, we found that, if we were not to presume (2.4), we would be left with the inequality (2.18). Investigating (2.18), we noticed that it would be sufficient to solve (2.18) for the controls u_i under some inner product. In the present section, we, in addition, allowed the inner product to vary smoothly with the position in state space, i.e. we allowed our state space to be a Riemannian manifold. The algebraic condition (2.18) on the controls from the previous section would here read

$$\exists \lambda > 0 : \quad \left\langle \Gamma_r^{\mathrm{id}} \pi, \sum_{i=1}^m g_i u_i \right\rangle \leq -\lambda V, \tag{2.69}$$

following from rewriting (2.56) in terms of the control vector fields. To solve this inequality for $u_1, \ldots u_m : M \to \mathbb{R}$ reveals the necessary and sufficient condition that, for all x in $U \setminus S$, the coefficients

$$\left\langle \Gamma_{r(x)}^x \pi(x), g_1(x) \right\rangle, \ \ldots \ , \left\langle \Gamma_{r(x)}^x \pi(x), g_m(x) \right\rangle, \tag{2.70}$$

which we denote by $c_1, \ldots, c_m : U \setminus S \to \mathbb{R}$ in the remainder of this chapter, are not all (simultaneously) zero. With the notation $c = (c_1, \ldots, c_m) : U \setminus S \to \mathbb{R}^m$ and $u = (u_1, \ldots, u_m) : M \to \mathbb{R}^m$, this is stated in the following lemma, which was published in [M3].

Lemma 1. There exist $\lambda > 0$ and $u : M \to \mathbb{R}^m$ (continuous away from S) such that, for all x in $U \setminus S$, (2.69) is satisfied, if and only if c is nonzero on $U \setminus S$.

Proof. We proof sufficiency and necessity separately.
if: First, find that $L_{g_1 u_1 + \cdots + g_m u_m} V$ equals $c \cdot u$. Since c does not attain the value zero on $U \setminus S$, let

$$u(x) = -\lambda \frac{V(x)}{c(x) \cdot c(x)} c(x) \tag{2.71}$$

for which $c \cdot u$ is $-\lambda V$. As a consequence, (2.69) remains satisfied for all x in $U \setminus S$.
only if: Should there exist $x \in U \setminus S$ with the property that $c(x) = 0$, then the left-hand side of (2.69), $c(x) \cdot u(x)$, is zero, while the right-hand side, $\lambda V(x)$ is nonzero. This proves the claim by contraposition. $\qquad\square$

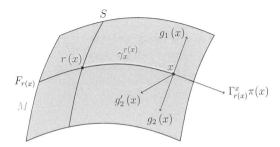

Figure 2.10: Orthogonality of control vector fields to $\Gamma_r^{\mathrm{id}}\pi$

As the inequality (2.69) is sufficient to render the Lie derivative of V along $f + g_1 u_1 + \cdots + g_m u_m$ negative away from S, it is a sufficient condition for asymptotic stabilizability of S through feedback. As we saw in the previous lemma, the coefficients $c_1, \ldots, c_m$ provide an algebraic characterization of the inequality (2.69). Together, we are now in the position to state an algebraic condition, which we previously published in [M3], for asymptotic stabilizability of S.

Proposition 3. Let U be a tubular neighborhood of S. If f is locally Lipschitz continuous and c is nonzero on $U \setminus S$, then there exists $u : M \to \mathbb{R}^m$, parametrized by a parameter k, such that, for every $\alpha > 0$ with $V^{-1}(\{\alpha\})$ contained in U, there exists $k_0 \geq 0$, such that for all $k > k_0$, S is an asymptotically stable invariant set of (2.1) and $V^{-1}([0, \alpha])$ is a subset of its region of asymptotic stability.

Proof. By linearity, the Lie derivative of V along $f + g_1 u_1 + \cdots + g_m u_m$ is given by

$$L_{f+g_1 u_1+\cdots+g_m u_m} V = L_f V + \sum_{i=1}^m L_{g_i u_i} V, \tag{2.72}$$

thus allowing us to investigate the Lie derivative of V along f separate from the Lie derivative of V along $g_1 u_1 + \cdots + g_m u_m$. The latter is given by

$$\sum_{i=1}^m L_{g_i u_i} V = c(x) \cdot u(x) \tag{2.73}$$

and application of Lemma 1 reveals that there exists $u : M \to \mathbb{R}^m$ such that $c \cdot u \leq -\lambda V$. Scaling this choice of u via $u \mapsto ku$ hence leads to the underestimate $c \cdot u \leq -k\lambda V$. Combining this result with the overestimate (2.53) for the Lie derivative of V along f leaves us with (2.57). The proof concludes alike the proof of Theorem 2. $\qquad\square$

We remark that the condition that c should not be zero away from S is equivalent to asking for the vector fields $g_1, \ldots, g_m$ to not be all (simultaneously) orthogonal to the vector field $\Gamma_r^{\mathrm{id}}\pi$ away from S, which we illustrated in Figure 2.10. Therein, the control vector fields g_1, g_2, depicted as blue arrows, are both orthogonal to $\Gamma_r^{\mathrm{id}}\pi$ (depicted as a blue arrow, as well) at x whilst the control vector fields g_1, g_2', indicated by blue arrows, are not (simultaneously) orthogonal to $\Gamma_r^{\mathrm{id}}\pi$ at this very point. This is, more geometrically, cast as requiring that application of the covector field $\Gamma_r^{\mathrm{id}}\pi^\flat$, obtained from lowering an index via the canonical (musical) isomorphism $\flat$, to the vector fields g_i yields (at least) one nonzero value at any point away from S. We further note that, as the vector field $\Gamma_r^{\mathrm{id}}\pi$ is tangent to the fibers $F_{r(x)}$, (one of which is plotted in black in Figure 2.10), if one equips the fibers F_x with the vector bundles NF_x, the condition that

$$\forall x \in U \setminus S, \quad \exists i: \quad g_i(x) \notin N_x F_{r(x)} \tag{2.74}$$

is necessary, but not sufficient, to guarantee that at least one of the vectors g_i will be nonorthogonal to $\Gamma_r^{\mathrm{id}}\pi$ away from S. The vector bundles NF_x, $x \in S$, should thus not contain all the vectors g_i at any point. Geometrically, this amounts to asking for the dimension of the linear combination of $N_x F_{r(x)}$ with the span of the vector fields $g_1, \ldots, g_m$ at x to be greater than the dimension of $N_x F_{r(x)}$, i.e. that for all x in $U \setminus S$,

$$\mathrm{span}\left(\{g_1(x), \ldots, g_m(x)\}\right) + N_x F_{r(x)} \neq N_x F_{r(x)}. \tag{2.75}$$

On the other hand, equipping the fibers F_x with the vector bundles TF_x, the condition that

$$\mathrm{span}\left(\{g_1(x), \ldots, g_m(x)\}\right) \supset T_x F_{r(x)}, \tag{2.76}$$

(cf. [24, section 4]) is sufficient, but not necessary, for c not to vanish.

Notably, the orthogonality condition (and also the directions of the normal and tangent spaces of the fibers F_x) depends on the choice of the Riemannian metric and, in fact, control vector fields which are (simultaneously) orthogonal to $\Gamma_r^{\mathrm{id}}\pi$ under one Riemannian metric may have a nonzero coefficient c_i under another Riemannian metric: for instance, if, at some x in $U \setminus S$, we have $\Gamma_{r(x)}^x\pi(x) = e_1$ and $g_i(x) = e_2$, presuming that $T_x M$ is $\mathbb{R}^2$, then the two are orthogonal under the metric tensor I_2, the 2×2 identity matrix, but not under the metric tensor $\left[\begin{smallmatrix} 2 & 1 \\ 1 & 2 \end{smallmatrix}\right]$. In other words, it is sufficient for asymptotic stabilizability of S if there exists a Riemannian metric $\langle \cdot, \cdot \rangle$ such that c is nonvanishing away from S. For M being $\mathbb{R}^n$, f being linear, g being constant, and S being the origin, finding such a metric tensor amounts to solving a Lyapunov equation.

Example 10 (Satellite Surveillance). In reconnaissance satellites, one must face a point on earth with a telescope. Modeling the satellite as a rigid body and letting the (body-fixed) telescope axis be e_1 (the first vector of the standard basis of $\mathbb{R}^3$), then this amounts to aligning xe_1 with another vector, say e_1 (without loss of generality), wherein x, a member of the special orthogonal group $\mathrm{SO}(3)$, describes the attitude of the satellite. In particular, all solutions in x to

$$xe_1 = e_1 \tag{2.77}$$

which belong to $\mathrm{SO}(3)$ solve the posed surveillance task. This problem statement is illustrated on the left-hand side of Figure 2.11, in which the point on earth to be faced by the telescope is indicated by a black circle, the telescope axis xe_1 is plotted as a black arrow, and the dashed line, connecting the satellite to the point on earth to be faced by the telescope, is spanned by the vector e_1, itself plotted as a black arrow. Parameterizing x with Euler angles, i.e.

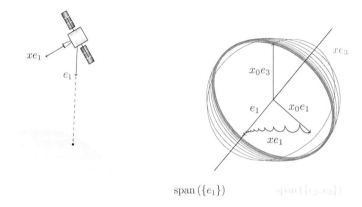

span $(\{e_1\})$

span $(\{e_2, e_3\})$

Figure 2.11: Satellite surveillance problem (left) and convergence of xe_1 to e_1 (right)

$$x = \text{rot}_1\,(\alpha_1)\,\text{rot}_3\,(\alpha_2)\,\text{rot}_1\,(\alpha_3) \tag{2.78}$$

$$= \begin{bmatrix} 1 & 0 & 0 \\ 0 & \cos(\alpha_1) & -\sin(\alpha_1) \\ 0 & \sin(\alpha_1) & \cos(\alpha_1) \end{bmatrix} \begin{bmatrix} \cos(\alpha_2) & -\sin(\alpha_2) & 0 \\ \sin(\alpha_2) & \cos(\alpha_2) & 0 \\ 0 & 0 & 1 \end{bmatrix} \begin{bmatrix} 1 & 0 & 0 \\ 0 & \cos(\alpha_3) & -\sin(\alpha_3) \\ 0 & \sin(\alpha_3) & \cos(\alpha_3) \end{bmatrix}$$

we find that (2.77) is equivalent to $\cos(\alpha_2) = 1$, i.e. $\text{rot}_3\,(\alpha_2) = I_3$, where I_3 denotes the identity of $\text{SO}\,(3)$. Any solution to (2.77) is thus parametrized as $x = \text{rot}_1\,(\alpha_1)\,\text{rot}_1\,(\alpha_3) = \text{rot}_1\,(\alpha_1 + \alpha_3)$, i.e. the homotopy circle $S = \text{rot}_1\,([0, 2\pi])$, a submanifold (and subgroup) of $\text{SO}\,(3)$ is the solution set of (2.77). Now, finding a controller that solves the surveillance problem amounts to bringing x towards S in a stable fashion. Suppose that the kinematic equations of the satellite are given by

$$\dot{x} = f\,(x) + g_1\,(x)\,u_1\,(x) + g_2\,(x)\,u_2\,(x) \tag{2.79}$$

where u_1, $u_2 : \text{SO}\,(3) \to \mathbb{R}$ are the controls sought, and f, g_1, g_2 are the drift and control vector fields

$$f\,(x) = x\Omega_1, \quad g_1\,(x) = x\Omega_2, \quad g_2\,(x) = x\Omega_3, \tag{2.80}$$

respectively, with Ω_1, Ω_2, Ω_3 being the infinitesimal generators of the Lie algebra $\mathfrak{so}\,(3)$. A drift in the kinematic equations could, for instance, be caused by solar (radiation) pressure. We emphasize that here, although S is reachable from a neighborhood of S (cf. [17]), the fact that f is nonvanishing on S but the control vector fields g_1, g_2 act only normal to S makes asymptotic stabilization of any singleton on S an impossible task, which has been extensively studied by Byrnes and Isidori [26]. However, S itself can be stabilized asymptotically, which is sufficient for solving the surveillance task. Applying the procedure proposed in this section, first, find that the restriction of

$$\phi\,(x) = -\frac{1}{2}\,\text{tr}\left(\log\left(\text{rot}_1\,(-\alpha_3)\,\text{rot}_3\,(-\alpha_2)\,\text{rot}_1\,(\alpha_3)\right)^2\right) \tag{2.81}$$

to the fibers F_x, $x \in S$, is strongly geodesically convex by construction; viz., the value of the function grows quadratically with the distance of its argument from S. The function ϕ is constructed by choosing a representative of x in S, $\mathrm{rot}_1(\alpha_1 + \alpha_3)$, and letting $\phi(x)$ grow with the distance of x to $\mathrm{rot}_1(\alpha_1 + \alpha_3)$, quantified by the distance of $x^\top \mathrm{rot}_1(\alpha_1 + \alpha_3) = \mathrm{rot}_1(-\alpha_3)\,\mathrm{rot}_3(-\alpha_2)\,\mathrm{rot}_1(\alpha_3)$ to the identity element I_3. Next, find that the gradient vector field $\mathrm{grad}\,\phi$ splits as a linear combination of g_1 and g_2, allowing us to solve (2.45) for $u = (u_1, u_2)$, yielding

$$u(x) = -k \begin{bmatrix} 0 & 0 & 1 \\ 0 & -1 & 0 \end{bmatrix} \log\left(\mathrm{rot}_1(-\alpha_3)\,\mathrm{rot}_3(-\alpha_2)\,\mathrm{rot}_1(\alpha_3)\right) e_1 \qquad (2.82)$$

with the convention that x and its Euler parameterization (2.78) agree. Here, S becomes an asymptotically stable invariant set for any k greater than $k_0 = 0$, as the Lie derivative of V along f is zero. For $k = 1$, a solution to (2.79) using the control (2.82) is plotted on the right-hand side of Figure 2.11, multiplied from the right by e_1 (blue plot) and e_3 (gray plot), respectively. As desired, xe_1 asymptotically aligns with e_1 (red arrow), whose span is plotted in black, while xe_3 approaches a periodic orbit contained in the linear hull of e_2 and e_3, indicated in light gray. The vectors resulting from multiplying e_1 and e_3 with the initial condition $x_0 = x(0)$ from the left are both illustrated by red arrows. We find that, here, although the surveillance tasked is solved by the control (2.82), the solutions contained in S are nontrivial.

The content of this example was previously published in [M3].

Example 11 (Controlling the Center of Mass through the Cross-Correlation). Consider the collection of dynamical systems

$$\dot{z}_i(t) = Az_i(t) + w_i(t) \qquad (2.83)$$

with i ranging from 1 to n. Suppose that we are in the position to decide upon $(w_1, \ldots, w_n) =: w$ but that n is too large to treat the (individual) signals w_i as decision variables, an issue arising in large-scale systems. Instead, one must opt to introduce an aggregate decision variable (cf. [3]), which is often chosen to be a statistical quantity of $(z_1, \ldots, z_n) =: z$. Here, we choose to aggregate the data w and z to the cross-correlation

$$(z_1, \ldots, z_n) \star (w_1, \ldots, w_n) = \frac{1}{n-1} \left(W - \bar{w} \otimes 1_n^\top\right)\left(Z - \bar{z} \otimes 1_n^\top\right)^\top \qquad (2.84)$$

and use it as decision variable, wherein 1_n denotes the n-fold vector of ones, W has the columns $w_1, \ldots, w_n$, Z has the columns $z_1, \ldots, z_n$, $\bar{z}$ is the center of mass of z and $\bar{w}$ is the center of mass of w. Notably, this decision variable can not be chosen arbitrarily: it must suffice the structural constraint imposed by the definition (2.84) of the cross-correlation. Suppose that our goal is to move the population z as a whole, i.e. to bring the center of mass $\bar{z}$ of z towards zero in a stable fashion. Moving a population as a whole, i.e. controlling its center of mass, is a control goal arising in large-scale systems (cf. [60]), for instance if the population is too large to control its members individually. We now turn our attention to the covariance of z. Remembering that the covariance of z is given by

$$\frac{1}{n-1}\left(Z - \bar{z} \otimes 1_n^\top\right)\left(Z - \bar{z} \otimes 1_n^\top\right)^\top =: x, \qquad (2.85)$$

we find that the evolution of the covariance x of z under the differential equations (2.83) is itself determined by the differential equation

$$\dot{x} = Ax + u(x) + xA^\top + u(x)^\top, \quad u : x \mapsto (z_1, \ldots, z_n) \star (w_1, \ldots, w_n) \tag{2.86}$$

wherein we restricted ourselves to computing our decision variable, the cross-correlation, solely from the covariance x (via the control u, still being sought) in order to bring the center of mass towards the origin, which amounts to bringing the covariance x, evolving on the convex cone of positive semidefinite matrices, M, towards its submanifold $S = \{x \text{ as in } (2.85) | \bar{z} = 0\}$ in a stable fashion. The controllability properties of (2.86) were extensively studied by Brockett [21, section 4], [23, section 10]. In order to asymptotically stabilize S, the function u can be obtained via the methods from this chapter, namely by first constructing a function ϕ which is strongly geodesically convex on the fibers F_x, $x \in S$, for instance

$$\phi(x) = \frac{1}{2}\left\| x - \frac{1}{n-1}ZZ^\top \right\|_\mathrm{F}^2, \tag{2.87}$$

which is chosen such that its value grows with the deviation of its argument from the zero center of mass covariance. Here, $\|\cdot\|_\mathrm{F}$ denotes the Frobenius norm. This construction is realized by choosing a representative of x in S, $\frac{1}{n-1}ZZ^\top$, and letting $\phi(x)$ grow with the distance of x to $\frac{1}{n-1}ZZ^\top$. Next, solving (2.45) for u yields

$$u(x) = -k\left(1_n^\top \otimes \bar{z} - \bar{z} \otimes 1_n^\top\right)\left(Z - \bar{z} \otimes 1_n^\top\right)^\top, \tag{2.88}$$

which stabilizes S asymptotically provided that k is greater than k_0, the largest eigenvalue of the symmetric part of A. It is quite remarkable that we implicitly insisted that the control u should be structured, i.e. that $u(x)$ must be a cross-correlation, but that our computation of u readily satisfied this structural constraint. When equating $u(x)$ with (2.84), we note that u is realized by $\bar{w} = -k\bar{z}$ and $W = -k1_n^\top \otimes \bar{z}$, i.e. by choosing each of the signals w_i to be $t \mapsto -k\bar{z}(t)$, which can be implemented even if one solely knows the center of mass of z, and which is a so-called broadcast control (cf. [6] or [43]) as it sends the same signal to each member of the population z. The connection between submanifold stabilization and structured controls will be strengthened in chapter 3.

The content of this example was previously published in [M3].

Asymptotic Stabilization of Equivalence Classes

In Proposition 3, we provided an algebraic characterization for asymptotic stabilizability, expressed in terms of the coefficients $c_i : U \setminus S \to \mathbb{R}$. We consequently discussed that the condition (2.75) is necessary, but not sufficient, for c to not vanish on $U \setminus S$. On the other hand, the stricter condition (2.76) is sufficient, but not necessary, for c to not vanish on $U \setminus S$. We already discussed how changing the Riemannian metric will tweak the direction of the normal and tangent spaces of the fibers F_x so as to alter the conditions (2.75) and (2.76). Another way to manipulate these conditions is by introducing an equivalence relation on M and by then analyzing the direction of the control vector fields g_i relative to the equivalence classes. Implicitly, we also did this with the conditions (2.75) and (2.76): as r uniquely retracts members of the tubular neighborhood onto S, we have that

$$x \in U \quad \Rightarrow \quad \exists! x' \in S : \quad x \in F_{x'} \tag{2.89}$$

and that hence, the relation $\sim$ on $U \times U$ defined by

$$x \sim x' \quad :\Leftrightarrow \quad \exists x'' \in S : \quad x, x' \in F_{x''} \tag{2.90}$$

is an equivalence relation with the equivalence class $[x]$ being the fiber $F_{x''}$. With this interpretation of (2.76) in mind, we next study the direction of the vector fields g_i relative to equivalence classes as a sufficient condition for asymptotic stabilizability not only for the equivalence relation (2.90), but also for more general equivalence relations. Choosing the equivalence relation with which we check the direction of the vector fields g_i relative to the equivalence classes provides degrees of freedom and simplification in verifying whether c is nonvanishing in a neighborhood of S, in order to apply Proposition 3. We, in particular, study equivalence relations under which S itself becomes an equivalence class; restricting ourselves to submanifolds S which are equivalence classes is motivated by problems such as the one from Example 10, wherein S is an equivalence class of the equivalence relation $x \sim x' :\Leftrightarrow \exists \alpha : x' = \mathrm{rot}_1(\alpha) x$. We first introduce a general setting for equivalence classes on $M \times M$ through quotient manifolds: let G be a Lie group and let $\mathsf{A} : \mathsf{G} \times M \to M$ be a proper, free, smooth, action of G on M. Then

$$x \sim x' \quad :\Leftrightarrow \quad \exists \mathsf{g} \in \mathsf{G} : \quad \mathsf{A}(\mathsf{g}, x) = x' \tag{2.91}$$

is an equivalence relation on $M \times M$ with the equivalence class $[x]$ being the orbit $\mathsf{A}(\mathsf{G}, x)$ of x under the action A. As A is proper, all equivalence classes are embedded submanifolds of M and we assume that $S \in M/\mathsf{G}$, where M/G denotes the quotient of A. Reconsidering Example 10 with this language, the Lie group $\mathsf{G} = \mathrm{rot}_1(\mathbb{R})$ and the left-multiplicative action $\mathsf{A} : (\mathsf{g}, x) \mapsto \mathsf{g}\, x$ bring the example to the present setup. In order to relate the orbits to the coefficients c_i, we equip the orbits M/G with fibers $V_x := T_x[x]$ and $H_x := N_x[x]$, whom we refer to as vertical and horizontal spaces, respectively. These vertical and horizontal spaces help us expressing and evaluating the direction of the vector fields g_i relative to the orbits. As A is free, it defines a foliation of M when treating its orbits as leaves. We say that a foliation is Riemannian if a geodesic that intersects orthogonally with a leave also intersects orthogonally with all other leaves that it meets. A submanifold of M is said to be totally geodesic when a geodesic on the submanifold is also a geodesic on M. The following proposition, which was previously published in [M3], relates the algebraic characterization of the previous proposition to horizontal and vertical spaces.

Proposition 4. Let U be a tubular neighborhood of S. If f is locally Lipschitz continuous and, for all x in $U \setminus S$, the linear hull of $g_i(x)$, $i = 1, \ldots, m$, contains H_x, then there exists a tubular neighborhood U' of S, contained in U, and a function $u : M \to \mathbb{R}^m$, parametrized by a parameter k, such that, for every $\alpha > 0$ with $V^{-1}(\{\alpha\})$ contained in U', there exists $k_0 \geq 0$, such that for all $k > k_0$, S is an asymptotically stable invariant set of (2.1) and $V^{-1}([0, \alpha])$ is a subset of its region of asymptotic stability. If, in addition, either (i) all orbits M/G are totally geodesic submanifolds of M, or (ii) the leaves M/G define a Riemannian foliation, then $U' = U$.

Proof. By definition, π has horizontal values. As A is a smooth action, horizontal spaces vary smoothly on M. Since, further, geodesics are smooth curves, for every x in U, there exists $\epsilon_x > 0$ such that $\Gamma_{r(x)}^{x'}\pi(x)$ is nonvertical, where x' is $\gamma_{r(x)}^x(\epsilon_x)$. Define

$$U' = \bigcup_{x \in U} \gamma_{r(x)}^x([0, \epsilon_x]) \tag{2.92}$$

to find that $\Gamma_r^{\mathrm{id}}\pi$ is nonvanishing and nonvertical on $U' \setminus S$. Hence, the orthogonal projection of $\Gamma_r^{\mathrm{id}}\pi$ onto horizontal spaces is nonzero on U'. For some $x \in U' \setminus S$, choose an orthonormal basis with vectors b_i for H_x and let ν_i denote the coordinates of the orthogonal projection of $\Gamma_{r(x)}^x\pi(x)$ onto H_x relative to the basis consisting of the vectors b_i. Pick a nonzero ν_i. Then $\langle \Gamma_{r(x)}^x\pi(x), \mathsf{b}_i \rangle = \nu_i$ is nonzero. But, as b_i is horizontal, it can be written as a linear combination of $g_i(x)$, $i = 1, \ldots, m$. Hence, there must be some nonzero $c_i(x)$. Application of Proposition 3, with U replaced by U', proves the first claim.

If (i) all orbits are totally geodesic, then the image of (x, V_x) under the exponential map is contained in $[x]$. However, the image of the composition of the exponential map with the section $(\mathrm{id}, -\Gamma_r^{\mathrm{id}}\pi)$ is contained in S. Hence, as distinct equivalence classes are disjoint, $\Gamma_r^{\mathrm{id}}\pi$ is nonvanishing and nonvertical on $U \setminus S$.

If (ii) the leaves M/G define a Riemannian foliation, then, as π is horizontal and nonvanishing on $U \setminus S$, the parallel transport $\Gamma_r^{\mathrm{id}}\pi$ is horizontal and nonvanishing on $U \setminus S$, as well. $\qquad\square$

The horizontal and vertical spaces provide degrees of freedom and simplification in verifying whether c is nonvanishing in a neighborhood of S. In fact, the previous proposition revealed that orbits of Lie group actions on M provided a more general framework for evaluating the use of the vector fields g_i than it was possible solely using the fibers F_x. Although the most general statement, allowing for arbitrary orbits, restricts the use of the proposition to a subneighborhood U' of U, conditions on the compatibility of the orbits with geodesics on M allowed to extend the applicability of the statement to U.

If the orbits are such that $V_x = N_x F_{r(x)}$, we recover the condition (2.76) from the condition

$$\mathrm{span}\left(\{g_1(x), \ldots, g_m(x)\}\right) \supset H_x \tag{2.93}$$

arising in the previous proposition. This being said, (2.76) is a special case of (2.93). On the other hand, as (2.76) asks for the linear hull of the vector fields g_i to contain certain tangent spaces whilst (2.93) asks for containing certain normal spaces, one could see the two conditions as being dual to each other.

The particular case in which the codimension of S in M, and hence the dimension of the horizontal spaces, is 1, deserves particular attention: for one coefficient c_i to not vanish, the condition (2.75) becomes necessary and sufficient in this case. In the foregoing proposition, for the vector fields g_i to contain H_x in their linear hull, it is necessary that at least one g_i is nonzero and horizontal. For the particular case that the leaves M/G define a Riemannian foliation, it becomes sufficient that at least one g_i is nonvertical (instead of the stronger condition that at least one g_i is nonzero and horizontal), as we saw in the proof that $\Gamma_r^{\mathrm{id}}\pi$ is horizontal and nonvanishing then.

Lastly, coming back to Example 10 once more, we find that the orbits $SO(3) / \mathrm{rot}_1 (\mathbb{R})$ are subgroups of $SO(3)$ whose Lie algebras are generated by Ω_1, each. Thus, the two control vector fields g_1, g_2 precisely span the horizontal spaces.

Example 12 (The Nonholonomic Integrator). Mechanical systems with nonintegrable constraints, so-called nonholonomic systems, have the reputation of being particularly hard to control (cf. [12, chapter 6] and references therein or [13] for a particular account on the relation of nonholonomic mechanical systems and asymptotic stabilization of submanifolds). The nonholonomic integrator

$$
\begin{aligned}
\dot{x}_1 &= v_1 (x) \\
\dot{x}_2 &= v_2 (x) \\
\dot{x}_3 &= x_1 v_2 (x) - x_2 v_1 (x)
\end{aligned}
\tag{2.94}
$$

is a prototype for a nonholonomic mechanical system for which no continuous control $v := (v_1, v_2)$ that stabilizes the origin asymptotically exists, although the system is controllable (cf. [19]). In [61], the control

$$
v(x) = \left(1 - x_1^2 - x_2^2\right) \begin{bmatrix} x_1 \\ x_2 \end{bmatrix} - u(x) \Omega \begin{bmatrix} x_1 \\ x_2 \end{bmatrix}
\tag{2.95}
$$

(for $u : x \mapsto 1$) was proposed, for which asymptotic stability of the unit cylinder $M = \{x \in \mathbb{R}^3 | x_1^2 + x_2^2 = 1\}$ can be verified. In fact, any function u which is uniformly bounded from below by a positive scalar stabilizes M asymptotically. Under the control v, M becomes an invariant set. Thus, if we ignore perturbations acting transversally to M and restrict our attention to initial condition on M, we may treat (2.94) under v as a system living on M. If we do so, the restriction that u must be bounded from below by a positive scalar becomes obsolete. The resulting system, restricted to M, reads

$$
\dot{x} = f(x) + g(x) u(x), \quad f(x) = -\Omega_3 x, \quad g(x) = e_3.
\tag{2.96}
$$

Suppose we aim to stabilize the submanifold $S = M \cap \mathrm{span}(\{e_1, e_2\})$ of M, the unit circle embedded in the unit cylinder, asymptotically, by choice of u. Then the method proposed above is applicable since f is tangent to S and S belongs to the quotient $M/\mathrm{rot}_3 (\mathbb{R})$ of the left multiplication $(g, x) \mapsto g\,x$. We find that g is a nonzero horizontal vector field and that all orbits $M/\mathrm{rot}_3 (\mathbb{R})$ are totally geodesic submanifolds of M. It follows from the foregoing proposition that a control $u : M \to \mathbb{R}$ can be found such that S is stabilized asymptotically. In particular, $u(x) = -kx_3$ turns S into an asymptotically stable invariant set for any k which is greater than $k_0 = 0$. In Figure 2.12, the unit cylinder M is depicted in black whilst its intersection with span $(\{e_1, e_2\})$, the unit circle, is depicted in blue. For some point x away from S, indicated by a red circle, its equivalence class $[x]$ and the control vector $g(x)$ are illustrated as a red line and a red arrow, respectively. We not only see that $[x]$ is a translated copy of S, but also that $g(x)$ is normal to $[x]$, making it a horizontal vector.

The content of this example was previously published in [M3].

Example 13 (Asymptotic Stabilization of the Equator). Spherical pendula and other mechanical systems involving spherical bearings or Cardan joints are examples of systems whose state spaces are the (unit) 2-sphere $\mathbb{S}_2 = M$, whom we treat as being

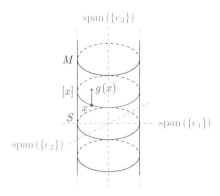

Figure 2.12: Control vector field and orbits for the nonholonomic integrator

embedded in $\mathbb{R}^3$. In this example, we consider the problem of bringing a system which lives on the 2-sphere towards its submanifold $S = \mathbb{S}_2 \cap \operatorname{span}(\{e_1, e_2\})$, i.e. the equator of the sphere, the unit circle embedded in the unit sphere. We consider the system

$$\dot{x} = f(x) + g_1(x) u_1(x) + g_2(x) u_2(x), \tag{2.97}$$

with drift and control vector fields

$$f(x) = (x_1 x_3 \Omega_2 - \Omega_3 - x_2 x_3 \Omega_1) x, \quad g_1(x) = -x_1 \Omega_2 x, \quad g_2(x) = x_3 \Omega_1 x \tag{2.98}$$

respectively, for whom a control $u := (u_1, u_2)$ is sought which asymptotically stabilizes the equator S. The three vector fields are depicted in Figure 2.13 as blue arrows, with the top plot depicting f, the middle plot depicting g_1, and the bottom plot depicting g_2. Notably, although f is tangent to S, S is an unstable invariant set of the unforced system (2.3). Further, we observe that both control vector fields vanish when approaching S on $\mathbb{S}_2 \cap \operatorname{span}(\{e_2, e_3\})$. In order to apply the method proposed above, we first need to endow $\mathbb{S}_2$ with an equivalence relation under whom S is an equivalence class. Here, as in the previous example, it is possible to choose the Lie group $G = \operatorname{rot}_3(\mathbb{R})$ and the action $A : (g, x) \mapsto g\,x$ since we find that S is the orbit of e_2 under A, plotted as a red line in Figure 2.13. Further, the orbits

$$\left[\operatorname{rot}_1\left(\frac{\pi}{16}\right)e_2\right], \quad \left[\operatorname{rot}_1\left(\frac{2\pi}{16}\right)e_2\right], \quad \left[\operatorname{rot}_1\left(\frac{3\pi}{16}\right)e_2\right], \quad \ldots \tag{2.99}$$

are depicted as black lines. We notice that the leaves $M/\,G$ define a Riemannian foliation and that there exists at least one nonvertical g_i in a neighborhood of S, allowing us to apply Proposition 3. Indeed, g_2 is nonvertical on $(\mathbb{S}_2 \setminus S) \cap \operatorname{span}(\{e_2, e_3\})$ while g_1 is nonvertical away from $\mathbb{S}_2 \cap \operatorname{span}(\{e_2, e_3\})$, implying that a control $u : \mathbb{S}_2 \to \mathbb{R}^2$ can be found which stabilizes the equator asymptotically. In fact,

$$u(x) = k \begin{bmatrix} 0 & 0 & 1 \\ 0 & 1 & 0 \end{bmatrix} x \tag{2.100}$$

turns S into an asymptotically stable invariant set provided that k exceeds $k_0 = 1$ by letting the system read

$$\dot{x} = (1 - k) f(x) - k\Omega_3 x, \qquad (2.101)$$

where $\dot{x} = -\Omega_3 x$ for $k = 1$ and $\dot{x} = -f(x) - 2\Omega_3 x$ for $k = 2$. Recalling the plot of f in Figure 2.13, it is evident that S must be an asymptotically stable invariant set of the reversed vector field $-f$.

The content of this example was previously published in [M3].

Summary

Starting from the previous section, we further investigated asymptotic stabilization of submanifolds for input affine systems such as (2.1), with the distinction that the state space of the system was a general Riemannian manifold in the present section. Most of the extensions discussed in the previous section are also valid in the present section but our focus was on overcoming the integrability assumption on $g_1 u_1 + \cdots + g_m u_m$. To do so, we derived an algebraic condition on the control vector fields, cast in terms of a vector field c, for which we asked not to vanish in a neighborhood of the submanifold. Last, under the assumption that S is an equivalence class, we were able to recast our algebraic condition for asymptotic stabilizability in terms of the direction of the vector fields g_i relative to equivalence classes close to S. We presented several examples for our approaches.

Discussion

In the last part of this section, we expressed conditions for asymptotic stabilizability in terms of the directions of the vector fields g_i relative to certain equivalence classes, of which we assumed S to be one. Conceptually similar, the idea of evaluating the directions of a vector field relative to equivalence classes was also employed by Forni and Sepulchre in [39, section VII] and [40, section VII] in order to specify limit sets of certain differential equations.

It shall be emphasized that the balancing problem, as formulated by Scardovi et al. [81], the formation control problem, as formulated by Sarlette et al. [77], the consensus optimization problem, as formulated by Sarlette and Sepulchre [78], as well as the distributed averaging problem, as formulated by Tron et al. [87], all fit well into the framework proposed in this section.

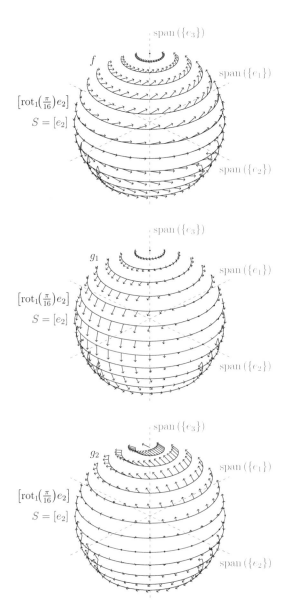

Figure 2.13: Asymptotic stabilization of the equator

3 Optimal Stabilization of Submanifolds

In the previous chapter, we proposed a method to find controls $x \mapsto u_i(x)$ for input affine systems of the form (2.1) which render a submanifold S of its state space asymptotically stable. Our method involved a parameter k which had to be chosen sufficiently large in order to guarantee asymptotic stability. Usually, controller design techniques of this kind will lead to large energy of the signals $t \mapsto u_i(x(t))$. Apart from asymptotic stability, one will thus have to worry about how to choose controls that have reasonably small energy but still bring solutions of (2.1) towards S, i.e. to find efficient solutions to the submanifold stabilization problem. To not a priori restrict ourselves to functions of the type $x \mapsto u_i(x)$ in our search for efficient controls, we treat the controls u_i as signals, i.e. as functions of the type $t \mapsto u_i(t)$ within this chapter, letting the input affine system (2.1) read

$$\dot{x}(t) = f(x(t)) + \sum_{i=1}^{m} g_i(x(t)) u_i(t), \quad x(0) = x_0. \tag{3.1}$$

In this chapter, we get back to the setting of section 2.1, i.e. the case where the state space of (3.1) is $\mathbb{R}^n$, meaning the drift vector field f and the control vector fields g_i are all $\mathbb{R}^n \to \mathbb{R}^n$ and S is a closed, smoothly embedded (in this chapter not necessarily compact) submanifold of $\mathbb{R}^n$. Throughout the chapter, we assume the drift and control vector fields to be continuously differentiable (even if not explicitly stated). We shall evaluate the efficiency of a control $u := (u_1, \ldots, u_m)$ via the value that the functional

$$(u : [0, T] \to \mathbb{R}^m) \quad \mapsto \quad \int_0^T d(x(t), S)^2 + u(t) \cdot Ru(t) \, \mathrm{d}t \tag{3.2}$$

attains under the given control u, where $T > 0$ and $x : [0, T] \to \mathbb{R}^n$ is related to $u : [0, T] \to \mathbb{R}^m$ via (3.1). Therein, $t \mapsto u(t) \cdot Ru(t)$ quantifies the energy of the controls, with R being a positive definite matrix subject to our choice, while $t \mapsto d(x(t), S)^2$ quantifies the proximity of the solution of (3.1) to the desired submanifold S of $\mathbb{R}^n$ (wherein $d(x, S)$ denotes the infimal Euclidean distance of x to all points in S). Asking for maintaining the functional (3.2) small amounts to finding a control that achieves a compromise between requiring small energy and bringing the solution of (3.1) towards S quickly. Among the measurable controls producing absolutely continuous solutions of (3.1) under which the functional (3.2) assumes a small value, we call those controls optimal, for which no other measurable controls producing absolutely continuous solutions of (3.1) yield smaller values for (3.2) (strictly speaking, we can thus only expect to satisfy (3.1) for almost all t but we will later find continuous optimal controls and thus ignore this subtle issue). Whenever we refer to an optimal control u and, in connection, refer to x, we shall refer to the solution of (3.1) under the optimal control u. The main results of this section were published in [M4].

This chapter is devoted to the analysis of such optimal controls. In particular, we investigate how S fixes a certain structure in optimal controls. Indeed, we find that all optimal controls u are state feedbacks, i.e. the definition of $t \mapsto u(t)$ explicitly involves the solution x of (3.1). We further find that the dependency of u on x is linear in $\pi \circ x$, with π as in (2.6), as long as the solution of (3.1) evolves in a tubular neighborhood U. More, the matrix defining this linear relationship, say K and $u : t \mapsto K(t) \pi(x(t))$, is structured in the sense that we know a subspace of its nullspace a priori, viz.

$$\forall t, \quad \ker(K(t)) \supset T_{r(x(t))} S. \tag{3.3}$$

Structured controls are of particular relevance in control theory (cf. [82, 74], wherein the structure of the controller is asked to agree with the structure of the control loop), as they admit for implementation based upon limited information (thus requiring fewer sensors), namely the information contained in the orthogonal complement of the nullspace of K. Further, by the rank-nullity theorem, a larger nullspace implies a smaller image; thus, structured controls will typically also admit implementation with limited actuation. In particular, applying the rank-nullity theorem to (3.3) reveals that the dimension of the image of K can not be larger than the codimension of S in $\mathbb{R}^n$. Hence, u may only range on a subspace of $\mathbb{R}^m$ whose dimension is smaller than or equal to the codimension of S in $\mathbb{R}^n$.

We will show that structured controls are necessary and, under additional assumptions, sufficient for optimality. Thus, the constructive aspect of our results is that the search for controls may a priori be restricted to structured controls. Further, the structure of the control will usually either simplify the equations determining the optimal control via a constructive Ansatz or directly reduce their dimension.

We study the optimal control problem of minimizing (3.2) subject to (3.1) in its Hamiltonian formalism. The Hamiltonian H of the problem is given by

$$H(x, u, a) = a \cdot f(x) + a \cdot \sum_{i=1}^{m} g_i(x) u_i - d(x, S)^2 - u \cdot Ru, \tag{3.4}$$

which satisfies the (generalized) Legendre-Clebsch condition. The third argument a of H is often referred to as the adjoint variable which must satisfy the differential equation

$$\dot{a} = -\frac{\partial}{\partial x} H(x, u, a) = -J_f(x) a - \sum_{i=1}^{m} J_{g_i}(x) au_i + 2\pi(x), \tag{3.5}$$

wherein $J_f : \mathbb{R}^n \to \mathbb{R}^{n \times n}$ denotes the Jacobian of f, subject to the boundary condition $a(T) = 0$ since the optimal control problem is Lagrangian (the partial derivative of H is calculated alike the gradient of V in section 2.1). The differential equation (3.5) is also referred to as adjoint equation of the Hamiltonian formalism. Letting $G : \mathbb{R}^n \to \mathbb{R}^{n \times m}$ have the columns g_i, the first-order necessary condition for optimality here yields the implication

$$\frac{\partial}{\partial u} H(x, u, a) = 0 \quad \Rightarrow \quad u = \frac{1}{2} R^{-1} G(x)^\top a \tag{3.6}$$

revealing that the optimal control u and the adjoint variable are indeed linearly related. Hence, in order to analyze the structure of u, it is sufficient to study the structure of a. We thus introduce the collections of all structured controls and all structured adjoint variables

$$\mathfrak{u}_x := \{u \in \mathbb{R}^m | \exists K_u \in \mathbb{R}^{m \times n} : \ker(K_u) \supset T_{r(x)}S \text{ and } u = K_u \pi(x)\} \qquad (3.7)$$

$$\mathfrak{a}_x := \{a \in \mathbb{R}^n | \exists \mathsf{P}_a \in \mathbb{R}^{n \times n} : \ker(\mathsf{P}_a) \supset T_{r(x)}S \text{ and } a = \mathsf{P}_a \pi(x)\} \qquad (3.8)$$

respectively, which are themselves real vector spaces for all x in U and hence define the vector bundles

$$\bigsqcup_{x \in U} \mathfrak{u}_x, \qquad \bigsqcup_{x \in U} \mathfrak{a}_x, \qquad (3.9)$$

over U. With this notation, asking whether u is structured amounts to asking whether u is in $\mathfrak{u}_x$. More geometrically, structured state feedbacks can be thought of as sections of the former of the two fiber bundles. Employing the notation P_a and K_u for a in $\mathfrak{a}_x$ and u in $\mathfrak{u}_x$ respectively, as introduced in (3.7), (3.8) above (we shall employ this notation without further ado in the remainder), we obtain that these two matrices are related to each other through

$$K_u = \frac{1}{2} R^{-1} G(x)^\top \mathsf{P}_a, \qquad (3.10)$$

via the first-order necessary condition (3.6). This reveals that, in order to verify that an optimal control is structured, it is sufficient to verify that a remains in $\mathfrak{a}_{x(t)}$ due to the inclusion $\ker(\mathsf{P}_a) \subset \ker(K_u)$. Our first result, which was published in [M4], states that $\mathfrak{a}_{x(t)}$ is indeed an invariant set of the adjoint equation, from which it follows that $T_{r(x)}S$ is an invariant subspace of the nullspace of P_a.

Lemma 2. Let u be an optimal control. For all t in $[0, T]$ such that $x(t)$ is in U, $\mathfrak{a}_{x(t)}$ is an invariant set of the adjoint equation (3.5).

Proof. As, for all x in U, $\mathfrak{a}_x$ is a real vector space, it is sufficient to prove the implication

$$a(t) \in \mathfrak{a}_{x(t)} \quad \Rightarrow \quad \dot{a}(t) \in \mathfrak{a}_{x(t)} \qquad (3.11)$$

(formally, first identify $T_{(x,a)} \bigsqcup \mathfrak{a}_x$ with $T_a \mathfrak{a}_x \oplus T_a \mathfrak{a}_x$). If $a(t)$ is in $\mathfrak{a}_{x(t)}$, then, using the representation $a = \mathsf{P}_a \pi(x)$ for $a \in \mathfrak{a}_x$, $\dot{a}(t)$ satisfies

$$\dot{a}(t) = \left(-\mathsf{J}_f(x(t)) \mathsf{P}_{a(t)} - \sum_{i=1}^{m} u_i(t) \mathsf{J}_{g_i}(x(t)) \mathsf{P}_{a(t)} + 2\mathsf{P}_{\pi(x(t))} \right) \pi(x(t)), \qquad (3.12)$$

according to the adjoint equation (3.5): thus $\mathsf{P}_{\dot{a}(t)}$ can be found such that $\ker(\mathsf{P}_{\dot{a}(t)})$ contains $T_{r(x(t))}S$ and $\dot{a}(t)$ equals $\mathsf{P}_{\dot{a}(t)} \pi(x(t))$, whence $\dot{a}(t)$ is in $\mathfrak{a}_{x(t)}$. $\qquad \square$

The function $U \to \mathbb{R}^{n \times n}$, $x \mapsto \mathsf{P}_{\pi(x)}$ occurred in the proof. As we will frequently encounter this function in the remainder of this chapter, we remark that $x' \mapsto \mathsf{P}_{\pi(x)} x'$ is precisely the orthogonal projection onto the normal space of S at $r(x)$.

With the boundary condition $a(T) = 0$ and the adjoint equation (3.5), we conclude that $\dot{a}(T) = 2\pi(x(T))$. Thus, we know that $a(t)$ must be in $\mathfrak{a}_{x(t)}$, i.e. be structured, as t approaches T. The previous lemma states that this structure in a is transported backwards, i.e. preserved, by the adjoint equation. This observation, together with the relationship (3.10), from which we know that the (invariant) structure of a is inherited by u, brings us into the position to state the main result of this chapter, which was previously published in [M4], namely that optimal controls (should they exist) are necessarily structured for all consecutive times for which x remains in a tubular neighborhood until time T.

Theorem 3. Let u be an optimal control. If $x(T)$ is in U, then, for all t in $(t_0, T]$, $u(t)$ is in $\mathfrak{u}_{x(t)}$, where t_0 is the infimal time among $[0, T]$ for which, for all t in $(t_0, T]$, $x(t)$ is in U.

Proof. As $\mathfrak{a}_{x(T)}$ is a vector space, the boundary condition $a(T) = 0$ implies that $a(T)$ is in $\mathfrak{a}_{x(T)}$. As, for all t in $(t_0, T]$, $x(t)$ is in U, Lemma 2 reveals that, for all t in $(t_0, T]$, $a(t)$ is in $\mathfrak{a}_{x(t)}$. The first-order optimality condition (3.6) reveals that the implication

$$a(t) \in \mathfrak{a}_{x(t)} \quad \Rightarrow \quad u(t) \in \mathfrak{u}_{x(t)} \tag{3.13}$$

is true, with K_u and P_a related via (3.10). This proves the claim. □

The restriction that $x(T)$ is contained in a tubular neighborhood of S appears little restrictive, for if the optimal control problem has a solution, then we expect that the optimal control brings the solution x close to S anyways. On the other hand, the validity of the statement of the theorem on the interval $(t_0, T]$ can be extended to the entire interval $[0, T]$ via the assumption that U is a forward invariant set of (3.1) under the optimal control and the initial condition $x(0)$ is contained in U. For simplicity, we impose this assumption in the remainder of this chapter. Not assuming forward invariance of U but instead asking for $x(T)$ to be in U and restricting ourselves to the interval $(t_0, T]$ such that, for all t in $(t_0, T]$, $x(t)$ is in U, will still be possible.

Example 14. In [20], Brockett formulates the pattern generation problem as bringing the solutions of

$$\dot{x} = f(x) + G(x)u, \quad f : \mathbb{R}^3 \to \mathbb{R}^3, \quad G : \mathbb{R}^3 \to \mathbb{R}^{3 \times 2} \tag{3.14}$$

towards one of its periodic orbits, where

$$f(x) = -\omega \Omega_3 x + e_3 \left(\omega \|\mathsf{P}_{12} x\|^2 - x_3 - 1 \right), \quad G(x) = \begin{bmatrix} 1 & 0 \\ 0 & 1 \\ -x_2 & x_1 \end{bmatrix}, \quad \mathsf{P}_{12} = \begin{bmatrix} 1 & 0 & 0 \\ 0 & 1 & 0 \end{bmatrix} \tag{3.15}$$

(Ω_3 is as in (2.67) and e_3 is the third vector of the standard basis of $\mathbb{R}^3$) for some $\omega > 0$ and the desired orbit is the submanifold $S = \{x \in \mathbb{R}^3 | \|\mathsf{P}_{12} x\|^2 = 1/\omega, x_3 = 0\}$ of $\mathbb{R}^3$ (we applied a few algebraic manipulation to bring the model from [20] to the above form). The control proposed by Brockett in [20] is $u = -x_3 \mathsf{P}_{12} x$. In this example, we briefly show that this control suffices the structural requirements which our previous theorem revealed to be necessary for optimality. In particular, $u = -x_3 \mathsf{P}_{12} x$ has the structured realization

$$u = - \begin{bmatrix} 0 & 0 & x_1 \\ 0 & 0 & x_2 \end{bmatrix} \pi(x), \quad \pi(x) = \mathsf{P}_{12}^\top \left(1 - \frac{\sqrt{1/\omega}}{\|\mathsf{P}_{12} x\|} \right) \mathsf{P}_{12} x + e_3 x_3, \tag{3.16}$$

where the inclusion $\ker(K) \supset T_{r(x)}S$ is verified by finding that $K\Omega_3 = 0$. For the given submanifold, in general, any optimal control $u = K\pi(x)$ must satisfy $K\Omega_3 x = 0$ in order to obtain $\ker(K) \supset T_{r(x)}S$, which is true if and only if

$$K_{11}x_2 = K_{12}x_1, \quad K_{21}x_2 = K_{22}x_1 \tag{3.17}$$

holds for the elements K_{ij} of K. The third column of K remains unrestricted. The content of this example was previously published in [M4].

Riccati Techniques

We did not yet require S to be an invariant set of the unforced system (2.3). This caused two shortcomings of the previous theorem: first, we could not yet show that $t \mapsto K(t)$ is continuous. Second, we could not yet let T tend to ∞. Both extensions will be discussed next, based upon the assumption that S is an invariant set of the unforced system (2.3), which we impose in the remainder of the chapter. The interpretation behind the significance of this assumption is that it enables controls which ultimately vanish as x approaches S, thus allowing for controls with finite energy even on infinite time intervals, independently of the precise location which x approaches on S. Should S, in contrast, not be an invariant set of (2.3), then it might, instead, be cheaper to move to a particular submanifold of S which is forward invariant under (3.1). In such cases, unstructured controls could be optimal (although the condition that the tangent space of the invariant submanifold of S at the retraction of x onto this invariant submanifold must be contained in the nullspace of K may still be necessary).

Should S be an invariant set of the unforced system (2.3), then, as f and π are continuously differentiable, (with a slight abuse of notation) we can find a continuous function of the form

$$\frac{\partial f}{\partial \pi} : U \to \mathbb{R}^{n \times n} \quad \text{such that} \quad \forall x \in U, \quad \mathsf{P}_{\pi(x)} f(x) = \frac{\partial f}{\partial \pi}(x) \pi(x), \tag{3.18}$$

i.e. a linear dependency of the orthogonal projection of $f(x)$ onto $N_{r(x)}S$ from $\pi(x)$. Since $\pi(x)$ is always in $N_{r(x)}S$, $\partial f/\partial \pi$ can be chosen such that its nullspace at x contains $T_{r(x)}S$. The notation $\partial f/\partial \pi$ is suggestive, for $\partial f/\partial \pi$ is the directional derivative of some function $(x, \pi(x)) \mapsto \mathsf{P}_{\pi(x)} f(x)$ along π (as such, it can be interpreted as a second order tensor field). Should S not be an invariant set of (2.3), then finding such a linear dependency is an impossible task as there will be some point x in S at which the orthogonal projection of $f(x)$ onto $N_{r(x)}S$ does not vanish although $\pi(x)$ does. The map $\partial f/\partial \pi$ is of particular use in the Riccati differential equation

$$\dot{X} = -X \frac{\partial f}{\partial \pi}(x) - \frac{\partial f}{\partial \pi}(x)^\top X - \mathsf{P}_{\pi(x)} + X\mathsf{P}_{\pi(x)} G(x) R^{-1} G(x)^\top \mathsf{P}_{\pi(x)} X \tag{3.19}$$

whose solution $X : [0, T] \to \mathbb{R}^{n \times n}$ determines the (as will be shown) optimal control

$$u = -R^{-1} G(x)^\top \mathsf{P}_{\pi(x)} X \pi(x), \tag{3.20}$$

thus providing a sufficient condition for optimality (in contrast to the necessary condition provided by the foregoing theorem). Here, as X solves a differential equation, it is smooth, hence letting the linear dependency $K = -R^{-1}G(x)^{\top} \mathsf{P}_{\pi(x)} X$ of u on π vary smoothly, as well. Our next proposition, which was published in [M4], not only claims optimality of the control (3.20), but also that X, and, as the inclusion $\ker(K) \supset \ker(X)$ remains true, also K, are indeed structured.

Proposition 5. If $u : [0,T] \to \mathbb{R}^m$ is as in (3.20), with $X : [0,T] \to \mathbb{R}^{n \times n}$ being the solution of the Riccati differential equation (3.19) subject to the boundary condition $X(T) = 0$, then u is an optimal control and, for all t in $[0,T]$, $\ker(X(t)) \supset T_{r(x(t))}S$.

Proof. Let J_π denote the Jacobian of π. As $\mathsf{J}_\pi = \mathsf{P}_\pi$, it follows from the chain rule that the differential equation governing $t \mapsto \pi(x(t))$ is

$$\dot{\pi}(x) = \frac{\partial f}{\partial \pi}(x)\,\pi(x) + \mathsf{P}_{\pi(x)}G(x)\,u \qquad (3.21)$$

with $\partial f / \partial \pi$ as in (3.18). Noting that the integrand $d(x,S)^2$ of (3.2) is nothing but $\pi(x) \cdot \mathsf{P}_{\pi(x)}\pi(x)$, it follows that $u : [0,T] \to \mathbb{R}^m$ as in (3.20), with $X : [0,T] \to \mathbb{R}^{n \times n}$ being the solution of the Riccati differential equation (3.19) subject to the boundary condition $X(T) = 0$ minimizes (3.2) subject to (3.1) (cf., e.g., [56, subsection 3.3.3]). It remains to show that, for all t in $[0,T]$, $\ker(X(t))$ contains $T_{r(x(t))}S$. Noting that the matrices X for which $\ker(X) \supset T_{r(x)}S$ constitute a real vector space, of which $X(T) = 0$ is a member, it is sufficient to prove the implication

$$\ker(X(t)) \supset T_{r(x(t))}S \quad \Rightarrow \quad \ker(\dot{X}(t)) \supset T_{r(x(t))}S \qquad (3.22)$$

(formally, again, first identify the double tangent bundle with direct sums of tangent spaces). To this end, assume that the nullspace of $X(t)$ contains $T_{r(x)}S$. Recalling that the nullspace of $\partial f / \partial \pi$ at x contains $T_{r(x)}S$, all expressions on the right-hand side of the Riccati differential equation (3.19) contain $T_{r(x)}S$ in their nullspaces, letting the implication (3.22) remain satisfied. This completes the proof. $\qquad\square$

It is important to note that the Riccati differential equation (3.19) can not be solved backwards from $X(T)$ to $X(0)$ in general as this requires explicit knowledge about the solution $x : [0,T] \to \mathbb{R}^n$ of (3.1), which, in turn, can not be found without knowing $X : [0,T] \to \mathbb{R}^{n \times n}$. Yet, it is still possible to infer structural properties of optimal controls from the Riccati differential equation (3.19) or to solve it via a structured Ansatz in particular cases, which will usually cause reduction of its dimension.

Example 15 (Synchronization). Consider the collection of systems

$$\dot{x}_i = A_0 x_i + B_0 u_i, \quad A_0 \in \mathbb{R}^{p \times p}, \quad i = 1, \ldots, n, \qquad (3.23)$$

whose solutions x_i we wish to synchronize, i.e. to bring the collective solution $x := (x_1, \ldots, x_n)$ towards the diagonal $\mathfrak{D}$ of $(\mathbb{R}^p)^n$ (which is $I_n \otimes A_0$-invariant), in an optimal fashion. Hence, we ask for minimization of the integral deviation of the individual solutions from their center of mass

$$\int_0^T \sum_{i=1}^n \left\| x_i(t) - \frac{1}{n} \sum_{j=1}^n x_j(t) \right\|^2 \mathrm{d}t \qquad (3.24)$$

plus the integral of $t \mapsto \sum_{i=1}^{n} u_i(t) \cdot R_0 u_i(t)$, over the interval $[0, T]$, for some positive definite matrix R_0. The Hamiltonian matrix associated with this optimal control problem is

$$\begin{bmatrix} I_n \otimes A_0 & -I_n \otimes B_0 R_0^{-1} B_0^\top \\ -\mathsf{P}_\pi & -I_n \otimes A_0^\top \end{bmatrix}, \quad \mathsf{P}_\pi = \left(I_n - \frac{1}{n} 1_n \otimes 1_n^\top \right) \otimes I_p, \qquad (3.25)$$

wherein $x \mapsto \mathsf{P}_\pi x$ is the orthogonal projection onto the orthogonal complement of $\mathscr{D}$. The associated Riccati differential equation reads

$$\dot{X} = -X \left(I_n \otimes A_0 \right) - \left(I_n \otimes A_0^\top \right) X - \mathsf{P}_\pi + X \left(I_n \otimes B_0 R_0^{-1} B_0^\top \right) X. \qquad (3.26)$$

By inspection, we notice that the solution of (3.26) is structured: if, as an Ansatz, we substitute $X = \left(I_n - \frac{1}{n} 1_n \otimes 1_n^\top \right) \otimes X_0$, then, should X_0 solve

$$\dot{X}_0 = -X_0 A_0 - A_0^\top X_0 - I_p + X_0 B_0 R_0^{-1} B_0^\top X_0, \qquad (3.27)$$

we obtain a solution of the Riccati differential equation (3.26). We note that the solution of the Riccati differential equation (3.27) assumes values in $\mathbb{R}^{p \times p}$ whilst the solution of the Riccati differential equation (3.26) lives in $\mathbb{R}^{np \times np}$, i.e. the structure in the solution of our optimal control problem allowed us to reduce the dimension n-fold. Once $X_0 : [0, T] \to \mathbb{R}^{p \times p}$ is obtained, the optimal control can be computed via $u = Kx$ where $K = \left(\frac{1}{n} 1_n \otimes 1_n^\top - I_n \right) \otimes R_0^{-1} B_0^\top X_0$ is structured since its nullspace is precisely the diagonal $\mathscr{D}$, making it a so-called diffusive coupling: every control can be written as the sum of differences

$$u_i = \frac{1}{n} \sum_{j=1}^{n} R_0^{-1} B_0^\top X_0 \left(x_j - x_i \right). \qquad (3.28)$$

These controls can be implemented using solely the relative information $x_j - x_i$ (which is relevant in applications involving relative sensing mechanisms).

A rather particular observation regarding the present example is that, should A_0 have all eigenvalues on the imaginary axis or in the open left half-plane, then, assuming the pair (A_0, B_0) to be controllable, initializing the solution of the Riccati differential equation (3.26) with P_π lets its solution converge to the strong (and thus maximal) solution of the associated algebraic Riccati equation, which can be established from the results of De Nicolao and Gevers [32, Theorem 3] / Callier and Willems [28, Theorems 3, and 4, subsection IV.C], respectively. This approach, as a byproduct, provides an optimal control for $T \to \infty$.

Borrelli and Keviczky [14] study the minimization of functionals similar to the ones considered in the present example, yet only asking for controls which asymptotically stabilize the origin. However, they arrive at distributed controllers in the sense that one merely needs to add a block-diagonal matrix to the matrix K obtained above in order to achieve asymptotic stability of the origin. Cao and Ren [30] study minimization of the above functional but restrict themselves to $A_0 = 0$ and $B_0 = I_p$. Fardad et al. [36] study the optimal synchronization problem from the present example in its $\mathcal{H}_2$ formulation, but impose the necessary structure obtained above, viz. that K contains the diagonal in its nullspace, as a constraint. Simultaneously, these authors maximize the amount of zero entries in K (its sparsity) via the reweighted ℓ^1 minimization proposed by Candès et al. [29].

The content of this example was previously published in [M18].

Example 16 (Broadcasting). Consider the collection of systems

$$\dot{x}_i = A_0 x_i + B_0 u_i, \quad A_0 \in \mathbb{R}^{p \times p}, \quad i = 1, \ldots, n, \tag{3.29}$$

which we want to move as a whole, i.e., control its center of mass, which amounts to bringing the solution $x := (x_1, \ldots, x_n)$ towards the orthogonal complement of the diagonal $\mathfrak{D}$ in $(\mathbb{R}^p)^n$ (which is $I_n \otimes A_0$-invariant), in an optimal fashion. Therefore, we ask for minimization of the integral of the center of mass

$$\int_0^T \left\| \frac{1}{n} \sum_{i=1}^n x_i(t) \right\|^2 \mathrm{d}t \tag{3.30}$$

plus the integral of $t \mapsto \sum_{i=1}^n u_i(t) \cdot R_0 u_i(t)$, over the interval $[0, T]$, for some positive definite matrix R_0.

We remark that this control problem is relevant in power networks if one should aim to produce the total power P_0 and the individual generators $1, \ldots n$ produce the powers $P_1, \ldots, P_n$. The change of variables $x_i = nP_i - P_0$ lets the sum of powers produced by the individual generators equal the desired power production P_0 if and only if the center of mass of $(x_1, \ldots, x_n)$ is at the origin.

Proceeding, the Hamiltonian matrix associated with the aforementioned optimal control problem is

$$\begin{bmatrix} I_n \otimes A_0 & -I_n \otimes B_0 R_0^{-1} B_0^\top \\ -\mathsf{P}_\pi & -I_n \otimes A_0^\top \end{bmatrix}, \quad \mathsf{P}_\pi = \frac{1}{n} 1_n \otimes 1_n^\top \otimes I_p, \tag{3.31}$$

wherein $x \mapsto \mathsf{P}_\pi x$ is the orthogonal projection onto the diagonal $\mathfrak{D}$ of $(\mathbb{R}^p)^n$. The associated Riccati differential equation reads as (3.26) (but with a different, viz. the above, P_π). By inspection, we notice that the solution of this Riccati differential equation is structured: if, as an Ansatz, we substitute $X = \frac{1}{n} 1_n \otimes 1_n^\top \otimes X_0$, then, should X_0 solve the Riccati differential equation (3.27), we obtain a solution of the present Riccati differential equation. We (again) note that the solution of the Riccati differential equation (3.27) assumes values in $\mathbb{R}^{p \times p}$ whilst the solution of our original Riccati differential equation lives in $\mathbb{R}^{np \times np}$, i.e. the structure in the solution of our optimal control problem allowed us to reduce the dimension n-fold. Once the solution $X_0 : [0, T] \to \mathbb{R}^{p \times p}$ of the Riccati differential equation (3.27) is obtained, the optimal control can be computed via $u = Kx$ where $K = -\frac{1}{n} 1_n \otimes 1_n^\top \otimes R_0^{-1} B_0^\top X_0$ is structured since its nullspace is precisely the orthogonal complement of the diagonal $\mathfrak{D}$, making it a so-called broadcast control (cf. [6] or [43]): every system receives the same signal

$$u_i = -R_0^{-1} B_0^\top X_0 \frac{1}{n} \sum_{i=1}^n x_i. \tag{3.32}$$

These controls can be transmitted to the individual systems without distinguishing between them in the transmission process. Further, they solely require computation of the center of mass and can thus be implemented using anonymized data such as the multiset $\{x_i\}_{i=1}^n$.

Madjidian and Mirkin [60] study the functional arising in this example (and also mention the relevance for power networks), yet only asking for controls which asymptotically stabilize the origin. However, they arrive at distributed controllers in the sense that one merely needs to add a block-diagonal matrix to the matrix K obtained above in order to achieve asymptotic stability of the origin. Asymptotic stabilization only using broadcast controls was discussed by Brockett [22].

The content of this example was previously published in [M4].

The Infinite Horizon Optimal Control Problem

We next consider the limiting case $T \to \infty$. When doing so, even if $\partial f / \partial \pi$ and $\mathsf{P}_\pi G$ were constant, the pair $(\partial f / \partial \pi, \mathsf{P}_\pi G)$ will in general not be controllable. This is because the codimension of S in $\mathbb{R}^n$, henceforth denoted by p, is, except for the trivial case of S being a singleton, smaller than n. To retain a (possibly) controllable pair, we have to reduce the dimension of the pair whose controllability we ask for from $(n \times n, n \times m)$ to $(p \times p, p \times m)$. This is done in the following fashion: let the columns of $\mathsf{E}_{\pi(x)} \in \mathbb{R}^{n \times p}$ be an orthonormal basis of $N_{r(x)}S$, i.e. $\mathsf{E}_\pi \mathsf{E}_\pi^\top = \mathsf{P}_\pi$. If there exist fixed (i.e. constant and, in particular, independent of x) $A \in \mathbb{R}^{p \times p}$ and $B \in \mathbb{R}^{p \times m}$ such that

$$A := \mathsf{E}_{\pi(x)}^\top \frac{\partial f}{\partial \pi}(x) \mathsf{E}_{\pi(x)}, \qquad B := \mathsf{E}_{\pi(x)}^\top G(x) \tag{3.33}$$

and should (A, B) be a controllable pair, then the unique positive definite solution $X_+ \in \mathbb{R}^{p \times p}$ of the algebraic Riccati equation

$$0 = -XA - A^\top X - I_p + XBR^{-1}BX \tag{3.34}$$

provides the structured optimal control

$$u = -R^{-1}B^\top X_+ \mathsf{E}_{\pi(x)}^\top \pi(x), \tag{3.35}$$

which is formalized in the following proposition that was originally published in [M4]. Here, $K = -R^{-1}B^\top X_+ \mathsf{E}_{\pi(x)}^\top$ is structured as the image of $\mathsf{E}_{\pi(x)}$ is $N_{r(x)}S$ and the nullspace of the transpose is just the orthogonal complement of the image, whence $\ker(K)$ contains the orthogonal complement of the image of $\mathsf{E}_{\pi(x)}$ in $\mathbb{R}^n$. Moreover, $K\mathsf{E}_{\pi(x)} = -R^{-1}B^\top X_+$ is constant for all x, i.e. apart from the expression E_π, the linear relationship between u and π is constant (independent of time).

Proposition 6. Let $T \to \infty$ and let the pair (A, B), defined as in (3.33), be controllable. If $u : [0, \infty) \to \mathbb{R}^m$ is as in (3.35), with $X_+ \in \mathbb{R}^{p \times p}$ being the unique positive definite solution of the algebraic Riccati equation (3.34), then u is an optimal control.

Proof. Consider the signal

$$t \mapsto \mathsf{E}_{\pi(x(t))}^\top \pi(x(t)), \tag{3.36}$$

whose evolution is, since the image of $\dot{\mathsf{E}}_{\pi(x)}$ is the tangent space of S at $r(x)$, governed by the differential equation

$$\dot{\mathsf{E}}_{\pi(x)}^\top \pi(x) + \mathsf{E}_{\pi(x)}^\top \dot{\pi}(x) = \mathsf{E}_{\pi(x)}^\top \left(\frac{\partial f}{\partial \pi}(x) \pi(x) + \mathsf{P}_{\pi(x)} G(x) u \right) \tag{3.37}$$

where we applied the product rule and substituted (3.21). Using $\pi = \mathsf{P}_\pi \pi$ and $\mathsf{P}_\pi = \mathsf{E}_\pi \mathsf{E}_\pi^\top$, the right-hand side of (3.37) simplifies to $A\mathsf{E}_\pi^\top \pi + Bu$. Noting that the integrand $d(x, S)^2$ of (3.2) is nothing but $\mathsf{E}_{\pi(x)}^\top \pi(x) \cdot I_p \mathsf{E}_{\pi(x)}^\top \pi(x)$, it follows that $u : [0, \infty) \to \mathbb{R}^m$ as in (3.35), with $X_+ \in \mathbb{R}^{p \times p}$ being the unique positive definite solution of the algebraic Riccati equation (3.34), minimizes (3.2) subject to (3.1) (cf., e.g., [56, subsection 3.4.3]). □

It shall be remarked that the structured optimal control (3.35), with X_+ the unique positive definite solution of the algebraic Riccati equation (3.34), indeed turns the origin of $\mathbb{R}^p$ into an asymptotically stable equilibrium of (3.37). For compact S, this will usually (provided that the optimal control renders a tubular neighborhood forward invariant plus a sublevel set of the Lyapunov function for (3.37) is contained in that tubular neighborhood) imply that S becomes an asymptotically stable invariant set of (3.1).

Although the assumption that constant matrices A, B, sufficing (3.33) can be found severely restricts the nonlinearity of f and G in the sense that their components lying in the normal spaces of S must look the same in any of these normal spaces, the components of f and G lying in the tangent spaces of S remain unrestricted.

Example 17. Consider the system

$$\dot{x} = \omega(x) \Omega x + \left(1 + \frac{1}{\|x\|} (u - 1) \right) x \tag{3.38}$$

with Ω as in (2.19) and $\omega : \mathbb{R}^2 \to \mathbb{R}$ arbitrary. Suppose some $u : [0, \infty) \to \mathbb{R}$ minimizing

$$\int_0^\infty d(x(t), \mathbb{S}_1)^2 + u(t)^2 \, dt \tag{3.39}$$

is sought, i.e. a control which brings the solution of (3.38) towards the unit circle $\mathbb{S}_1 = \{ x \in \mathbb{R}^2 | \, \|x\| = 1 \}$, embedded in $\mathbb{R}^2$, with as little energy as possible. Here, using

$$\pi(x) = x - \frac{1}{\|x\|} x, \quad \mathsf{P}_{\pi(x)} = \frac{1}{x \cdot x} x \otimes x^\top, \quad \mathsf{E}_{\pi(x)} = \frac{1}{\|x\|} x, \tag{3.40}$$

we have that $\mathsf{P}_\pi f = \pi$, hence letting $\partial f / \partial \pi$ be the identity matrix I_2 and furthermore revealing that $A = 1$ via (3.33). On the other hand, we find that $B = \mathsf{E}_\pi^\top G = 1$, yielding the controllable pair $(A, B) = (1, 1)$. Recalling that the codimension of $\mathbb{S}_1$ in $\mathbb{R}^2$ is 1, the algebraic Riccati equation (3.34) thus becomes

$$0 = -2X - 1 + X^2, \tag{3.41}$$

whose unique positive solution is $X_+ = 1 + \sqrt{2}$, from which we compute the structured optimal control

$$u = \left(1 + \sqrt{2} \right) \left(1 - \|x\| \right) \tag{3.42}$$

according to (3.35). Under this optimal control, our system reads

$$\dot{x} = \omega(x) \Omega x - \sqrt{2} \pi(x), \tag{3.43}$$

of whom $\mathbb{S}_1$ is an asymptotically stable invariant set (as can be verified using the Lyapunov function from section 2.1).

The content of this example was previously published in [M4].

Numerical Integration

In applications of optimal control, one often fixes some (small) $T > 0$ to then initialize the adjoint equation (3.5) with $a(T) = 0$ in order to (numerically) integrate it backwards until obtaining $a : [0, T] \to \mathbb{R}^n$. Consequently, $u : [0, T] \to \mathbb{R}^m$ as in (3.6) is applied to the system (3.1) without intermediate measurements of x; this procedure is repeated recurrently. If one was to apply such a procedure in submanifold stabilization problems, it would be desirable to maintain the structure of the optimal control, which Theorem 3 revealed to be necessary for optimality anyways, despite numerical errors arising during integration of the adjoint equation. This can be realized in the following fashion: the matrices

$$\Lambda_{ij}(x) = e_i \otimes e_j^\top \mathsf{E}_{\pi(x)}^\top, \quad e_i \in \mathbb{R}^n, \quad e_j \in \mathbb{R}^p, \quad i = 1, \ldots, n, \quad j = 1, \ldots, p \qquad (3.44)$$

form an orthonormal basis of the subspace of $\mathbb{R}^{n \times n}$ consisting of the matrices X for which $\ker(X) \supset T_{r(x)}S$. Hence, the vectors $\Lambda_{ij}\pi$ contain $\mathfrak{a}_x$ inside their linear hull. Thus, by Moore-Penrose, the linear map

$$a \mapsto \sum_{i=1}^{n} \sum_{j=1}^{p} \left(\Lambda_{ij}(x) \frac{1}{\|\pi(x)\|} \pi(x) \right) \left(\Lambda_{ij}(x) \frac{1}{\|\pi(x)\|} \pi(x) \right)^\top a \qquad (3.45)$$

is the best approximation of a in $\mathfrak{a}_x$ and the matrix defining the linear relationship between a and its best approximation in $\mathfrak{a}_x$ is the best approximation of P_a among the matrices containing $T_{r(x)}S$ in their nullspace, allowing us to compute a structured K_u according to (3.10).

Summary

In this chapter, we studied the problem of bringing the solution of a system towards a desired submanifold in an optimal fashion, i.e. with as little control energy as possible. We determined necessary (Theorem 3) and sufficient (Propositions 5 and 6) conditions for optimality, but all conditions involved the structure of the optimal control. In particular, all our controls solely involved knowledge about the map π (which first appeared in section 2.1) instead of knowledge about the entire state of the system, hence allowing to implement these controls with incomplete sensing. Moreover, the dependency of the optimal control on π was found to be linear, with the matrix defining this linear relationship containing the tangent spaces of the desired submanifold in its nullspace. This, by the rank-nullity theorem, implies that the optimal control can not attain arbitrary values and may thus, in practice, be realized using limited actuation. We discussed how structured adjoint equations could be integrated numerically and presented several examples of structured optimal controls.

Discussion

For $S = \{0\}$, linear f, and constant G, our optimal control problem becomes the linear-quadratic regulator problem (cf., e.g., [56]). As such, our optimal controls suggested in Propositions 5 and 6 could be seen as a natural extension of the linear-quadratic regulator to submanifold stabilization problems. This is best reflected by the relevance of the linear-quadratic regulator for the proofs of Propositions 5 and 6. In drawing this analogy, it must be remarked that the control proposed in Proposition 5 relies upon a Riccati differential equation whose solution depends on x_0. Studying the proof of Proposition 6 further, one finds that, if the measurement $x \mapsto C\mathsf{E}_{\pi(x)}\pi(x)$ was available for computation of the control (instead of the value of $\pi(x)$ itself), wherein the pair (C, A) is expected to be observable, then it is possible to uniquely reconstruct $\pi(x_0)$ (and hence $\pi(x)$) from these measurements, say via an appropriate observer, in order to compute and apply the proposed optimal control. In doing so, the observer may be designed independently of the control suggested above (cf., e.g., [50, section 4.2]).

Scherer [82] as well as Qi et al. [74] study structured optimal / robust controllers (in the $\mathcal{H}_\infty$- and combined ℓ^1, $\mathcal{H}_2$, $\mathcal{H}_\infty$-sense, respectively) for linear systems, admitting for rather general zero-patterns in K (imposed as a constraint), which is formally (yet not conceptually) quite different from our condition on the nullspace of K. In a similar vein, Katzberg [52] has asked for minimization of quadratic functionals whilst constraining K to obey given zero-patterns. For positive linear systems, $\mathcal{H}_\infty$ solutions with arbitrary zero patters were constructed by Tanaka and Langbort [86] via structured versions of the Kalman-Yakubovich-Popov and bounded real lemmata.

In this chapter, the solution of (3.1) is forced towards S by letting the integrand of the functional (3.2), which we ask to minimize, vanish only if the solution of (3.1) is contained in S. An alternative approach would be to impose an endpoint constraint which restricts the controls amongst which we seek for optimal controls to those which let the solution of (3.1) ultimately reach S. For S being a subspace of $\mathbb{R}^n$, this direction was taken by Soethoudt and Trentelman [85], yet with the focus on conditions for existence of optimal controls rather than their structure.

Employing only the limited information provided by $\mathsf{E}_{\pi(x)}^\top \pi(x)$ for computation of the control u, as we did in Proposition 6, can be seen as aggregation of the data x. The concept of data aggregation was brought to the (optimal) control community by Aoki [3] for application to linear large-scale systems, who showed that the suboptimality of controllers solely employing aggregated data (thus being structured) can be bounded from above, provided that the functional to be minimized is quadratic. In contrast, we showed above that controllers which solely employ the aggregated data $\mathsf{E}_{\pi(x)}^\top \pi(x)$ (instead of the original data x) are indeed optimal for minimization of the functional (3.2) whose integrand only depends (quadratically) on the aggregated data itself. Examples 15 and 16 explicitly applied these findings to large-scale systems.

Similar to the interests pursued in this chapter (particularly Examples 15 and 16), Ho and Chu [48] studied the structure of optimal controls in team decision problems by assigning a graph to the information pattern which realizes the optimal control.

4 Input-Output Methods for Submanifold Stabilization

In the foregoing chapter, we asked for the functional (3.2) to become minimal through our choice of control. In this chapter, we relax this requirement and merely ask for a functional of the form (3.2) to remain finite even if we let T tend to ∞. Meanwhile, we consider a more general system model.

In particular, a system will in this chapter be a relation which relates an input signal to an output signal. This type of models is particularly relevant in applications for which no first principles modeling with differential equations is possible (for instance biology, sociology, or economics) or state-space models are not available for other reasons (for instance when the system is too complex to describe it in closed form). On the other hand, input-output relations might be a more suitable model when the underlying system actually produces output data for given inputs, such as experimental setups or numerical models do. For general references regarding the input-output approach, we refer to the seminal work of Zames [97, 98] or the monograph by Desoer and Vidyasagar [33], to both of which our approach is conceptually very similar, though these references only study setpoint regulation problems. The main results of this section were published in [M6].

More particularly, should x be the quantity of interest, assuming values in $\mathbb{R}^n$, which we wish to bring towards the closed, smoothly embedded (in this chapter not necessarily compact) submanifold S of $\mathbb{R}^n$ through the control u, then our system is the relation H_1 which relates $u : t \mapsto u(t)$ to $x : t \mapsto x(t)$, i.e. $(u, x) \in H_1$ (systems of the form (3.1) fall well inside this model). On the other hand, the controller is the relation H_2 which relates x to u again, i.e. $(x, u) \in H_2$. In addition, we allow for the input of the system H_1 to be perturbed by some exogenous signal $t \mapsto w(t)$ in the form that $e := u + w$ is the input of H_1, such that indeed $(e, x) \in H_1$, leading to the feedback interconnection depicted in Figure 4.1. Therein, one calls H_1 the feedforward path and H_2 the feedback path.

In the spirit of the functional (3.2) whose minimization we studied in the previous chapter, we will be concerned with the question whether the integral p-fold distance of

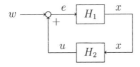

Figure 4.1: Feedback interconnection of system H_1 and controller H_2

x to S as well as the integral p-fold norm of u remain finite, i.e. whether

$$\int_{\mathbb{R}} d\left(x\left(t\right),S\right)^{p} \mathrm{d}\,t < \infty, \qquad \int_{\mathbb{R}} \left\|u\left(t\right)\right\|^{p} \mathrm{d}\,t < \infty. \tag{4.1}$$

Here, by $\int \mathrm{d}\,t$, we mean the Lebesgue integral (presuming that both x and u are measurable functions). We remark that it is possible to replace $\mathbb{R}$ by $[0,\infty)$ throughout. Asking whether these two inequalities are fulfilled amounts to asking whether u and x are contained in

$$\mathscr{L}^{p,m} := \left\{ \Phi : \mathbb{R} \to \mathbb{R}^{m} \,\middle|\, \Phi \text{ is a measurable function and } \int_{\mathbb{R}} \left\|\Phi\left(t\right)\right\|^{p} \mathrm{d}\,t \text{ is finite} \right\} \tag{4.2}$$

$$\mathscr{L}_{S}^{p,n} := \left\{ \Phi : \mathbb{R} \to U \,\middle|\, \Phi \text{ is a measurable function and } \int_{\mathbb{R}} d\left(\Phi\left(t\right),S\right)^{p} \mathrm{d}\,t \text{ is finite} \right\} \tag{4.3}$$

respectively, whereby U is a given tubular neighborhood of S to which we again have to restrict our attention as we will see further below. With measurable functions, we refer to Lebesgue measurable functions. We remark that $\Phi : \mathbb{R} \to U$ could be replaced by $\Phi : \mathbb{R} \to \mathbb{R}^{n}$ in the last definition if we would require that $\Phi\left(t\right) \in U$ for almost all t in $\mathbb{R}$. It shall be noted that $\mathbb{R}^{n}$ is a tubular neighborhood of its origin $\{0\}$, whence $\mathscr{L}_{\{0\}}^{p,n} = \mathscr{L}^{p,n}$.

Example 18. Let $\mathbb{S}_{1}$ be the unit circle, embedded in $\mathbb{R}^{2}$, and replace the domain $\mathbb{R}$ in (4.2), (4.3) by $[0,\infty)$ for simplicity. Consider the family of signals

$$\Phi : t \mapsto \begin{bmatrix} \left(1 + a\left(t\right)\right)\cos\left(t\right) \\ \left(1 + a\left(t\right)\right)\sin\left(t\right) \end{bmatrix} \tag{4.4}$$

with $\left|a\left(t\right)\right|$ smaller than 1 for all t, such that Φ maps to a tubular neighborhood of $\mathbb{S}_{1}$. If a is a member of $\mathscr{L}^{2,1}$, for instance $a\left(t\right) = \alpha\frac{\sin(t)}{1+t}$, then Φ is in $\mathscr{L}_{\mathbb{S}_{1}}^{2,2}$. If, in contrast, a is not in $\mathscr{L}^{2,1}$, say $a\left(t\right) = \left(\alpha - \epsilon\right)\sin\left(t\right)$ for some small but positive ϵ, then Φ is also not in $\mathscr{L}_{\mathbb{S}_{1}}^{2,2}$. Both signals are depicted in Figure 4.2, with the $\mathscr{L}_{\mathbb{S}_{1}}^{2,2}$ signal on the left and the signal which is not in $\mathscr{L}_{\mathbb{S}_{1}}^{2,2}$ on the right. The tubular neighborhood U is depicted in gray and the circle $\mathbb{S}_{1}$ in black – the signals are plotted in blue. The $\mathscr{L}_{\mathbb{S}_{1}}^{2,2}$ signal decays to the circle but the signal which is not in $\mathscr{L}_{\mathbb{S}_{1}}^{2,2}$ crosses the circle periodically such that its distance to $\mathbb{S}_{1}$ never vanishes and thus has unbounded integral.

As we do not a priori want to assume that H_{1} is a relation on $\mathscr{L}^{p,m} \times \mathscr{L}_{S}^{p,n}$ and H_{2} is a relation on $\mathscr{L}_{S}^{p,n} \times \mathscr{L}^{p,m}$ (which would in fact trivialize the control problem), we next define the extensions of $\mathscr{L}^{p,m}$ and $\mathscr{L}_{S}^{p,n}$. Extensions are supersets of $\mathscr{L}^{p,m}$ and $\mathscr{L}_{S}^{p,n}$ which also contain signals that are in $\mathscr{L}^{p,m}$ or $\mathscr{L}_{S}^{p,n}$ when sent to zero or to S (respectively) for all times greater than or equal to t', for all real t'. Sending signals to zero or to S is done by the truncation operator $\Phi \mapsto \Phi_{S}^{t'}$, which, for some $\Phi : \mathbb{R} \to U$ returns the signal $\Phi_{S}^{t'} : \mathbb{R} \to U$, defined by

$$\Phi_{S}^{t'} : t \mapsto \begin{cases} \Phi\left(t\right) & \text{for } t < t' \\ \left(r \circ \Phi\right)\left(t\right) & \text{elsewhere} \end{cases} \tag{4.5}$$

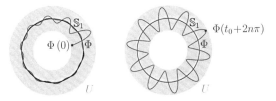

Figure 4.2: Signals which are (left) or are not (right) in $\mathcal{L}_{\mathbb{S}_1}^{2,2}$

wherein r is the (smooth) retraction onto S defined just as in section 2.1. We remark that this truncation might also be applied for $S = \{0\}$ such that $\Phi_{\{0\}}^{t'}$ assumes the value zero for all arguments larger than or equal to t', which is just the classical truncation operator (cf., e.g., [33, chapter III, section 1]). Alike the notation $\mathcal{L}_{\{0\}}^{p,n} = \mathcal{L}^{p,n}$, we also write $\Phi^{t'}$ instead of $\Phi_{\{0\}}^{t'}$. Having the truncation operator at hand, we define the extension

$$\bar{\mathcal{L}}_S^{p,n} := \left\{ \left. \Phi : \mathbb{R} \to U \; \right| \text{ for all } t' \text{ in } \mathbb{R}, \Phi_S^{t'} \text{ is in } \mathcal{L}_S^{p,n} \right\} \tag{4.6}$$

of $\mathcal{L}_S^{p,n}$. This also defines the (usual) extension $\bar{\mathcal{L}}^{p,m}$ of $\mathcal{L}^{p,m}$.

Example 19. Reconsider the signal Φ from Example 18 with $\mathbb{S}_1$ being the unit circle, embedded in $\mathbb{R}^2$ and $a(t) = (\alpha - \epsilon) \sin(t)$ for some small but positive ϵ. Here, the truncation of Φ reads

$$\Phi_{\mathbb{S}_1}^{t'} : t \mapsto \begin{cases} \begin{bmatrix} (1 + a(t)) \cos(t) \\ (1 + a(t)) \sin(t) \end{bmatrix} & \text{for } t < t' \\[2ex] \begin{bmatrix} \cos(t) \\ \sin(t) \end{bmatrix} & \text{elsewhere} \end{cases} \tag{4.7}$$

where we used $r : x \mapsto (1/\|x\|) x$. As $a^{t'}$ is in $\mathcal{L}^{2,1}$ for any real t', it follows that Φ is in the extension $\bar{\mathcal{L}}_{\mathbb{S}_1}^{2,2}$. In Figure 4.3, we depict Φ (left) as well as $\Phi_{\mathbb{S}_1}^{t'}$ (right) for some t'. The tubular neighborhood is again depicted in gray and the circle in black. As we recall from the foregoing example, the left signal never reached $\mathbb{S}_1$ for good. On the other hand, the truncated signal remains on the circle for all arguments greater than or equal to t' but does not come to rest, i.e. it oscillates around the origin with radius 1. From the figure, we infer that the signal is sent to the circle along the normal space of $\mathbb{S}_1$ at $(r \circ \Phi)(t')$ (depicted in red), or more general along the fiber $F_{(r \circ \Phi)(t')}$, as the truncation was defined via the retraction from the tubular neighborhood onto $\mathbb{S}_1$.

The content of this example was previously published in [M6].

Using this notation, the submanifold stabilization problem can be cast as follows: given a subset H_1 of $\bar{\mathcal{L}}^{p,m} \times \bar{\mathcal{L}}_S^{p,n}$, find a subset H_2 of $\bar{\mathcal{L}}_S^{p,n} \times \bar{\mathcal{L}}^{p,m}$ such that, for any w in $\mathcal{L}^{p,m}$, x is in $\mathcal{L}_S^{p,n}$ and, if possible, u is in $\mathcal{L}^{p,m}$, where w, x, and u are related to

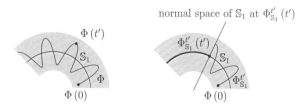

Figure 4.3: Signal which is not in $\mathscr{L}^{2,2}_{\mathbb{S}_1}$ (left), but whose truncation (right) is

each other through the feedback equations

$$w + u = e, \tag{4.8}$$

$$(e, x) \in H_1 \subset \bar{\mathscr{L}}^{p,m} \times \bar{\mathscr{L}}^{p,n}_S, \tag{4.9}$$

$$(x, u) \in H_2 \subset \bar{\mathscr{L}}^{p,n}_S \times \bar{\mathscr{L}}^{p,m}. \tag{4.10}$$

The first requirement that x is in $\mathscr{L}^{p,n}_S$ translates to x becoming asymptotically close to S, quantified with the p-fold distance. The second, optional, requirement that u is in $\mathscr{L}^{p,m}$ translates to employing only finite energy for solving this task. Introducing the relations E_u and E_x which relate w to e and w to x, respectively, subject to the feedback equations (4.8)-(4.10), and calling a relation on $\bar{\mathscr{L}}^{p,n}_S \times \bar{\mathscr{L}}^{p,m}_{S'}$ bounded (wherein S is a submanifold of $\mathbb{R}^n$ and S' is a submanifold of $\mathbb{R}^m$) if it should relate members of $\mathscr{L}^{p,n}_S$ only to members of $\mathscr{L}^{p,m}_{S'}$, then our control problem is to find conditions under which E_x and, if possible, E_u, becomes bounded. An important tool to evaluate whether a function belongs to some given $\mathscr{L}^{p,n}_S$ will be the functional

$$\|\cdot\|_{\mathscr{L}^{p,n}_S} : \mathscr{L}^{p,n}_S \to \mathbb{R}, \quad \Phi \mapsto \left(\int_{\mathbb{R}} d\left(\Phi\left(t\right), S\right)^p \mathrm{d}\,t \right)^{1/p} \tag{4.11}$$

as $\|\Phi\|_{\mathscr{L}^{p,n}_S}$ is finite whenever Φ is in $\mathscr{L}^{p,n}_S$.

One is tempted to think that defining $\|\cdot\|_{\mathscr{L}^{p,n}_S}$ is sufficient to apply classical techniques such as the small-gain theorem, the feedback theorem for conic relations, or the feedback theorem for passive relations [97, 98]. However, the proofs of these results and even the definitions of conicity and passivity rely on the fact that $(\mathscr{L}^{p,n}, \|\cdot\|_{\mathscr{L}^{p,n}})$ are normed vector spaces which are complete in the norm $\|\cdot\|_{\mathscr{L}^{p,n}}$, i.e. that they are Banach spaces and that $(\mathscr{L}^{2,n}, \langle\cdot, \cdot\rangle_{\mathscr{L}^{2,n}})$ with

$$\langle\cdot, \cdot\rangle_{\mathscr{L}^{2,n}} : \mathscr{L}^{2,n} \times \mathscr{L}^{2,n} \to \mathbb{R}, \quad (\Phi, \Phi') \mapsto \int_{\mathbb{R}} \Phi\left(t\right) \cdot \Phi'\left(t\right) \mathrm{d}\,t \tag{4.12}$$

is even a Hilbert space. Further, the proofs of these results involve the usual inequalities (triangle inequality, Cauchy-Schwarz, etc.). Unfortunately, $\left(\mathscr{L}^{p,n}_S, \|\cdot\|_{\mathscr{L}^{p,n}_S}\right)$ is not complete, nor normed, nor merely vector space.

Example 20. Let $\mathbb{S}_1$ be the unit circle, embedded in $\mathbb{R}^2$. Consider the signals $\Phi : t \mapsto -e_i$ and $\Phi' : t \mapsto (1 + \epsilon) e_i$ for some positive but small ϵ (e_i is a vector of the standard

Figure 4.4: $\mathscr{L}_{\mathbb{S}_1}^{\infty,2}$ violates the triangle inequality (left) and is no vector space (right)

basis of $\mathbb{R}^2$). Then, as $\Phi(t) \in \mathbb{S}_1$, $d(\Phi'(t), S) = \epsilon$, and $\Phi + \Phi' : t \mapsto \epsilon e_i$, we have

$$\|\Phi\|_{\mathscr{L}_{\mathbb{S}_1}^{\infty,2}} + \|\Phi'\|_{\mathscr{L}_{\mathbb{S}_1}^{\infty,2}} = \epsilon, \quad \|\Phi + \Phi'\|_{\mathscr{L}_{\mathbb{S}_1}^{\infty,2}} = 1 - \epsilon \qquad (4.13)$$

wherein $1 - \epsilon$ exceeds ϵ for small enough ϵ, thus violating the triangle inequality. Furthermore, considering the limiting case $\epsilon = 0$, we find that $\Phi + \Phi'$ maps any real argument to the origin and thus assumes values outside any tubular neighborhood of $\mathbb{S}_1$ (the origin is not contained in any tubular neighborhood of $\mathbb{S}_1$), hence revealing that $\mathscr{L}_{\mathbb{S}_1}^{\infty,2}$ is no vector space. All signals are depicted in Figure 4.4 as blue circles (this is possible as all signals have singleton images). On the left, ϵ is chosen small and we infer that $\Phi + \Phi'$ is indeed farther away from $\mathbb{S}_1$, depicted in black, than Φ'. On the right, ϵ is sent to zero and we see that $\Phi + \Phi'$ maps to the origin, which lies outside the tubular neighborhood U, drawn in gray.

The content of this example was previously published in [M6].

The Small-Gain Theorem

Although $\|\cdot\|_{\mathscr{L}_S^{p,n}}$ does not satisfy the usual inequalities, it can be used to define a meaningful (induced) operator "norm": for given $\mathscr{L}_S^{p,n}$ and $\mathscr{L}_{S'}^{p,m}$, with S being a submanifold of $\mathbb{R}^n$ and S' a submanifold of $\mathbb{R}^m$, the function

$$\text{gain} : H \mapsto \sup \left\{ \left. \frac{\|\Phi_{S'}'^t\|_{\mathscr{L}_{S'}^{p,m}}}{\|\Phi_S^t\|_{\mathscr{L}_S^{p,n}}} \right| (\Phi, \Phi') \in H, \ t \in \mathbb{R}, \ \|\Phi_S^t\|_{\mathscr{L}_S^{p,n}} \neq 0 \right\} \qquad (4.14)$$

from the power set of $\mathscr{L}_S^{p,n} \times \mathscr{L}_{S'}^{p,m}$ to $\mathbb{R}$ will turn out to be useful in verifying whether some relation H on $\mathscr{L}_S^{p,n} \times \mathscr{L}_{S'}^{p,m}$ is bounded. In particular, H is bounded if $\text{gain}(H)$ is bounded (although the converse is not true) and the overestimate

$$\|\Phi_{S'}'^t\|_{\mathscr{L}_{S'}^{p,m}} \leq \text{gain}(H) \|\Phi_S^t\|_{\mathscr{L}_S^{p,n}} \qquad (4.15)$$

remains true for any Φ' which is related to Φ through H. Using this function, we are able to state a generalization of the small-gain theorem, which was previously published in [M6].

Lemma 3. Let E_u and E_x relate w to e and x, respectively, subject to (4.8)-(4.10). If the product of $\text{gain}(H_1)$ and $\text{gain}(H_2)$ is smaller than 1, then E_u and E_x are bounded.

Proof. Consider any w, e, and u solving the feedback equations. Truncation yields $w^t + u^t = e^t$ for any real t. Noting that w^t, u^t, and e^t are all in $\mathscr{L}^{p,m}$, we infer that the $\mathscr{L}^{p,m}$-norm of e^t is smaller than or equal to the sum of the $\mathscr{L}^{p,m}$-norms of w^t and u^t from the triangle inequality. Application of (4.15) thus yields

$$\|e^t\|_{\mathscr{L}^{p,m}} \leq \text{gain}(H_2) \|x_S^t\|_{\mathscr{L}_S^{p,n}} + \|w^t\|_{\mathscr{L}^{p,m}} \qquad (4.16)$$

and applying (4.15) once more, we arrive at

$$\left\|e^t\right\|_{\mathscr{L}^{p,m}} \leq \text{gain}\,(H_1)\,\text{gain}\,(H_2)\left\|e^t\right\|_{\mathscr{L}^{p,m}} + \left\|w^t\right\|_{\mathscr{L}^{p,m}} \tag{4.17}$$

from which, under the condition that the product of $\text{gain}\,(H_1)$ and $\text{gain}\,(H_2)$ is smaller than 1, it follows that

$$\left\|e^t\right\|_{\mathscr{L}^{p,m}} \leq \frac{1}{1 - \text{gain}\,(H_1)\,\text{gain}\,(H_2)}\left\|w^t\right\|_{\mathscr{L}^{p,m}}. \tag{4.18}$$

Now, should w be in $\mathscr{L}^{p,m}$, then letting t tend to ∞ provides a real upper bound for the $\mathscr{L}^{p,m}$-norm of e, proving that E_u is bounded. It remains to show that E_x is bounded. To do so, consider any x solving the feedback equations. Truncating and applying (4.15) yields

$$\left\|x_S^t\right\|_{\mathscr{L}_S^{p,n}} \leq \text{gain}\,(H_1)\left\|u^t\right\|_{\mathscr{L}^{p,m}} \tag{4.19}$$

and substituting the feedback equations with subsequent application of (4.15), we arrive at

$$\left\|x_S^t\right\|_{\mathscr{L}_S^{p,n}} \leq \text{gain}\,(H_1)\left\|w^t\right\|_{\mathscr{L}^{p,m}} + \text{gain}\,(H_1)\,\text{gain}\,(H_2)\left\|x_S^t\right\|_{\mathscr{L}_S^{p,n}} \tag{4.20}$$

from which, under the condition that the product of $\text{gain}\,(H_1)$ and $\text{gain}\,(H_2)$ is smaller than 1, it follows that

$$\left\|x_S^t\right\|_{\mathscr{L}_S^{p,n}} \leq \frac{\text{gain}\,(H_1)}{1 - \text{gain}\,(H_1)\,\text{gain}\,(H_2)}\left\|w^t\right\|_{\mathscr{L}^{p,m}}. \tag{4.21}$$

Now, should w be in $\mathscr{L}^{p,m}$, then letting t tend to ∞ provides a real upper bound for the value which the functional $\|\cdot\|_{\mathscr{L}_S^{p,n}}$ assumes at x, proving that E_x is bounded. This was the last statement to be proven. $\qquad\square$

Example 21. Let H_1 relate e to

$$x : t \mapsto \frac{\|e\,(t)\| + 1}{\|e\,(t)\|}e\,(t) \tag{4.22}$$

and let H_2 relate x to

$$u : t \mapsto k\frac{\|x\,(t)\| - 1}{\|x\,(t)\|}x\,(t) \tag{4.23}$$

for some positive scalar k. Should S be the unit circle $\mathbb{S}_1$, embedded in $\mathbb{R}^2$, then $\text{gain}\,(H_1)$ is 1 and $\text{gain}\,(H_2)$ is precisely k such that our foregoing lemma tells us to choose k smaller than 1, i.e. somewhere in the open unit interval $(0,1)$. If we iteratively compute x and e as above for $w = 0$, which can be thought of as introducing a sample-and-hold operator into the feedback interconnection, we arrive at the iteration

$$x_{i+1} = \left(k + \frac{1}{\|\pi\,(x_i)\|}\right)\pi\,(x_i) \tag{4.24}$$

(with π as in (2.6)) which generates the Cauchy sequence

$$\frac{1}{2}e_1, \quad -\frac{5}{4}e_1, \quad -\frac{9}{8}e_1, \quad -\frac{17}{16}e_1, \quad \ldots, \quad -\frac{2^i + 1}{2^i}e_1, \quad \ldots \tag{4.25}$$

for $x_0 = \frac{1}{2}e_1$ and $k = \frac{1}{2}$, whose subsequence $\left(-\frac{2^{i+1}+1}{2^{i+1}}e_1 \right)_{i \in \mathbb{N}}$ converges to $-e_1 \in \mathbb{S}_1$, implying that it converges to $-e_1$ itself. This iterative process of solving the feedback equations bridges the gap from $\mathscr{L}^{p,n}$ to the sequence spaces $\ell^{p,n}$, or, more general,

$$\ell_S^{p,n} = \left\{ \Phi : \mathbb{N} \to U \,\middle|\, \sum_{i=1}^{\infty} d\left(\Phi\left(i\right), S\right)^p < \infty \right\}, \tag{4.26}$$

for which the results in this chapter hold, as well (formally by replacing $\mathbb{R}$ with $\mathbb{N}$ and the Lebesgue measure on $\mathbb{R}$ with the counting measure on $\mathbb{N}$).

The content of this example was previously published in [M6].

The condition of the foregoing lemma, viz. that the product of $\text{gain}\,(H_1)$ and $\text{gain}\,(H_2)$ is smaller than 1, appears to be seldom met in practice (according to [97]). However, it is possible to find transformed relations H_1' and H_2', the product of whose gains is significantly smaller than the product of the gains of the original relations H_1 and H_2. In particular, the product of the gains of the transformed relations may be smaller than 1 even if the product of the gains of the original relations is larger than or equal to 1. Thereby, we will construct the transformed relations in such a fashion, so as to guarantee that they will be contained in certain cones. Definition of these cones will require an inner product with induced norm, i.e. we will have to work in some Hilbert space. We recall that $\mathscr{L}_S^{2,n}$ with $\|\cdot\|_{\mathscr{L}_S^{2,n}}$ is not even a vector space. The next lemma, which was published in [M6], states that the composition operator $\pi\circ$, with π defined as in section 2.1, allows us to work with members of $\mathscr{L}_S^{2,n}$ as if they were members of the Banach space $\mathscr{L}^{p,n}$.

Lemma 4. For any Φ in $\bar{\mathscr{L}}_S^{p,n}$, for any t in $\mathbb{R}$, $\pi \circ \Phi$ is in $\bar{\mathscr{L}}^{p,n}$ and the identities

$$\left(\pi \circ \Phi\right)^t = \pi \circ \Phi_S^t := \pi \circ \left(\Phi_S^t\right), \quad \left\|\pi \circ \Phi_S^t\right\|_{\mathscr{L}^{p,n}} = \left\|\Phi_S^t\right\|_{\mathscr{L}_S^{p,n}} \tag{4.27}$$

hold true. If, moreover, Φ is in $\mathscr{L}_S^{p,n}$, then $\pi \circ \Phi$ is in $\mathscr{L}^{p,n}$.

Proof. If Φ is in $\bar{\mathscr{L}}_S^{p,n}$, then we have

$$\int_{\mathbb{R}} d\left(\Phi_S^t\left(t'\right), S\right)^p \mathrm{d}\,t' = \int_{-\infty}^{t} d\left(\Phi\left(t'\right), S\right)^p \mathrm{d}\,t' < \infty \tag{4.28}$$

for any real t. Using $d\left(x, S\right) = \|\pi\left(x\right)\|$, this also reveals that

$$\int_{-\infty}^{t} \left\|\left(\pi \circ \Phi\right)\left(t'\right)\right\|^p \mathrm{d}\,t' = \int_{\mathbb{R}} \left\|\left(\pi \circ \Phi\right)^t\left(t'\right)\right\|^p \mathrm{d}\,t < \infty \tag{4.29}$$

for any real t, whence $\pi \circ \Phi$ is in $\bar{\mathscr{L}}^{p,n}$ and $\left\|\pi \circ \Phi_S^t\right\|_{\mathscr{L}^{p,n}} = \left\|\Phi_S^t\right\|_{\mathscr{L}_S^{p,n}}$. Omitting the truncation, this also proves the implication

$$\Phi \text{ is in } \mathscr{L}_S^{p,n} \quad \Rightarrow \quad \pi \circ \Phi \text{ is in } \mathscr{L}^{p,n}. \tag{4.30}$$

It remains to show that $\left(\pi \circ \Phi\right)^t = \pi \circ \Phi_S^t$. Recalling the definition of the truncation $\Phi \mapsto \Phi_S^t$, we find that the restriction of $\pi \circ \Phi_S^t$ to $[t, \infty)$ is precisely the restriction of $r \circ \Phi - r \circ r \circ \Phi$ to $[t, \infty)$. Using the identity $r \circ r = r$, this proves our last claim. $\square$

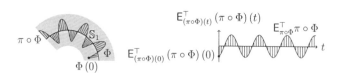

Figure 4.5: A signal from $\bar{\mathscr{L}}^{2,2}_{\mathbb{S}_1}$ (left) and its representation in $\bar{\mathscr{L}}^{2,1}$ (right)

With the knowledge of the previous lemma, we may indeed refer to the composition operator $\pi\circ$ as the map

$$\pi\circ:\bar{\mathscr{L}}^{p,n}_S\to\bar{\mathscr{L}}^{p,n},\quad\Phi\mapsto\pi\circ\Phi,\quad\pi\circ(\mathscr{L}^{p,n}_S)\subset\mathscr{L}^{p,n},\tag{4.31}$$

which we do in the remainder of the chapter, where $\pi\circ(\mathscr{L}^{p,n}_S)$ is to be understood as the image of $\mathscr{L}^{p,n}_S$ under composition with π.

Example 22. Reconsider the signal Φ from Example 18 with $\mathbb{S}_1$ being the unit circle, embedded in $\mathbb{R}^2$ and $a(t)=(\alpha-\epsilon)\sin(t)$ for some small but positive ϵ. As the component of Φ lying in the fibers $F_{r\circ\Phi}$ (the fibers being defined as in section 2.1) is precisely a, we have that $\mathsf{E}^\top_{\pi\circ\Phi}\pi\circ\Phi=a$, with $\mathsf{E}_{\pi(x)}$ a matrix whose columns form an orthonormal basis of $N_{r(x)}S$ as in the previous chapter. In Figure 4.5, we depict Φ (left) as well as the representation of $\pi\circ\Phi$ in the normal spaces of $\mathbb{S}_1$, i.e. $\mathsf{E}^\top_{\pi\circ\Phi}\pi\circ\Phi$ (right), both in blue. On the left, we see that $\pi\circ\Phi$, plotted in red, attains values in the normals spaces of $\mathbb{S}_1$ at $r\circ\Phi$. Representing it in the basis E_π (on the right), unrolls the signal to the real line, where it appears as the signal $t\mapsto\sin(t)$, scaled by $\alpha-\epsilon$.

We invite keen readers to compare the figure to [92, Fig. 3], wherein the integral distance of a solution of a linear system to a subspace is depicted.

The content of this example was previously published in [M6].

Conic Relations

Henceforth, let $p=2$ and $m=n$.

With $p=2$ and the above properties of the map $\pi\circ$ at hand, for two submanifolds S and S' of $\mathbb{R}^n$, we are in the position to define the conic relations

$$\mathbb{K}^{\mathsf{r}}_{\mathsf{c}}:=\Big\{H\subset\bar{\mathscr{L}}^{2,n}_S\times\bar{\mathscr{L}}^{2,n}_{S'}\Big|\text{ for all }(\Phi,\Phi')\text{ in }H,\text{ for all }t\text{ in }\mathbb{R},\tag{4.32}$$
$$\big\langle\pi'\circ\Phi''_{S'}-(\mathsf{c}-\mathsf{r})\,\pi\circ\Phi^t_S,\pi'\circ\Phi''_{S'}-(\mathsf{c}+\mathsf{r})\,\pi\circ\Phi^t_S\big\rangle_{\mathscr{L}^{2,n}}\leq0\Big\}$$
$$\bar{\mathbb{K}}^{\mathsf{r}}_{\mathsf{c}}:=\Big\{H\subset\bar{\mathscr{L}}^{2,n}_S\times\bar{\mathscr{L}}^{2,n}_{S'}\Big|\text{ for all }(\Phi,\Phi')\text{ in }H,\text{ for all }t\text{ in }\mathbb{R},\tag{4.33}$$
$$\big\langle\pi'\circ\Phi''_{S'}-(\mathsf{c}-\mathsf{r})\,\pi\circ\Phi^t_S,\pi'\circ\Phi''_{S'}-(\mathsf{c}+\mathsf{r})\,\pi\circ\Phi^t_S\big\rangle_{\mathscr{L}^{2,n}}\geq0\Big\}$$

where $\mathsf{c}\in\mathbb{R}$ represents the center of the cone and $\mathsf{r}\geq0$ can be interpreted as the radius of the cone. Strictly speaking, one would have to distinguish between two different maps $\pi\circ$ and $\pi'\circ$ in the foregoing definitions since $\pi=\mathsf{P}_2\circ\mu^{-1}:U\to\mathbb{R}^n$ with $\mu:NS\to\mathbb{R}^n$, $(x,v)\mapsto x+v$, $\mathsf{P}_2:NS\to\mathbb{R}^n$, $(x,v)\mapsto v$ and U a tubular neighborhood of S was applied to $\Phi^t_S\in\mathscr{L}^{2,n}_S$ whilst $\pi'=\mathsf{P}_2\circ\mu^{-1}:U'\to\mathbb{R}^n$ with $\mu:NS'\to\mathbb{R}^n$,

$(x, v) \mapsto x + v$, $\mathsf{P}_2 : NS' \to \mathbb{R}^n$, $(x, v) \mapsto v$ and U' a tubular neighborhood of S' was applied to $\Phi_{S'}^t \in \mathscr{L}_{S'}^{2,n}$. We will, however, not explicitly distinguish between these two distinct usages of π in the remainder as its particular meaning can always be inferred from its argument: either S or S' will always be the origin whence either $\pi \circ$ or $\pi' \circ$ is the identity. The notation $\overline{\mathbb{K}}_c^r$ is suggestive since the union $\mathbb{K}_c^r \cup \overline{\mathbb{K}}_c^r$ is the entire power set of $\mathscr{L}_S^{2,n} \times \mathscr{L}_{S'}^{2,n}$.

We will refer to relations contained in the cones $\mathbb{K}_c^r$ or $\overline{\mathbb{K}}_c^r$ as conic relations. One main reason for considering conic relations is expressed in the following lemma, which was previously published in [M6].

Lemma 5. If H is in $\mathbb{K}_0^r$, then gain (H) is smaller than or equal to r.

Proof. Let H be contained in $\mathscr{L}_S^{2,n} \times \mathscr{L}_{S'}^{2,n}$. Consider some Φ' related to Φ through H. Exploiting the bilinearity of $\langle \cdot, \cdot \rangle_{\mathscr{L}^{2,n}}$, we arrive at

$$\left\| \pi \circ \Phi_{S'}'^t \right\|_{\mathscr{L}^{2,n}}^2 - r^2 \left\| \pi \circ \Phi_S^t \right\|_{\mathscr{L}^{2,n}}^2 \leq 0. \tag{4.34}$$

Substituting the identity $\left\| \pi \circ \Phi_S^t \right\|_{\mathscr{L}^{p,n}} = \left\| \Phi_S^t \right\|_{\mathscr{L}_S^{p,n}}$ from Lemma 4 and taking the square root, we have

$$\left\| \Phi_{S'}'^t \right\|_{\mathscr{L}_{S'}^{2,n}} \leq r \left\| \Phi_S^t \right\|_{\mathscr{L}_S^{2,n}}. \tag{4.35}$$

Taking the supremum over all related (Φ, Φ') and all real t, the claim is proven. $\quad\square$

We next introduce the transformations H_1', H_2' on $\mathscr{L}^{2,n} \times \mathscr{L}_S^{2,n}$ and $\mathscr{L}_S^{2,n} \times \mathscr{L}^{2,n}$, respectively, via

$$H_1' := \left(H_1^{-1} + k\pi \circ \right)^{-1} \tag{4.36}$$
$$H_2' := H_2 + k\pi \circ \tag{4.37}$$

for some real scalar k which yet needs to be determined (strictly speaking, $\pi \circ$ here has to be defined as the relation which relates Φ to $\pi \circ \Phi$). The transformed feedback equations, accordingly, read

$$w + u' = e', \tag{4.38}$$
$$(e', x') \in H_1', \tag{4.39}$$
$$(x', u') \in H_2', \tag{4.40}$$

and the relations E_u' and E_x' relate w to e' and w to x', respectively, subject to these transformed feedback equations (4.38)-(4.40). The feedback interconnection resulting from the above transformed feedback equations is depicted in Figure 4.6.

The reasons for defining the transformed feedback equations in this fashion are the overestimates provided by the following lemma, which was previously published in [M6].

Lemma 6. Let E_u, E_x relate w to e, x, respectively, subject to (4.8)-(4.10) and let E_u', E_x' relate w to e', x', respectively, subject to (4.38)-(4.40). The relations E_u and E_x are contained in the relations E_u' and E_x', respectively.

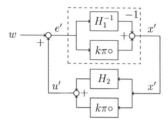

Figure 4.6: Transformed feedback interconnection

Proof. Consider some x and e solving the feedback equations. Should e' solve the transformed feedback equations, then, by the very definition of H_1', $e + k\pi \circ x$ is related to x through H_1'. Now consider some u solving the feedback equations and u' solving the transformed feedback equations. By the definition of H_2', x is related to $u + k\pi \circ x$ through H_2'. Lastly, finding that $e + k\pi \circ x = u + w + k\pi \circ x = u' + w$, we recover the transformed feedback equations. □

With this lemma at hand, we are in the position to state the main result of this chapter, which was previously published in [M6].

Theorem 4. Let E_u and E_x relate w to e and x, respectively, subject to (4.8)-(4.10). For all $r > 0$ and $c < r$, there exist c' in $\mathbb{R}$ and $r' > 0$ such that, for all H_1 in $\mathbb{K}_c^r$, for all $-H_2$ in $\mathbb{K}_{c'}^{r'}$, E_u and E_x are bounded. If $c > r$, the statement holds for all H_1 in $\bar{\mathbb{K}}_c^r$.

Proof. Define

$$c' := \frac{c}{r^2 - c^2}, \quad r' := \frac{r}{|r^2 - c^2|} \tag{4.41}$$

and let the parameter k from the transformed relations H_1' and H_2' be equal to c'. Choose some positive ϵ smaller than r'. Then, by some algebraic manipulations, we find that

$$-H_2 \in \mathbb{K}_{c'}^{r'-\epsilon} \quad \Rightarrow \quad H_2 \in \mathbb{K}_{-c'}^{r'-\epsilon} \quad \Rightarrow \quad H_2' \in \mathbb{K}_0^{r'-\epsilon} \tag{4.42}$$

whence gain (H_2') is smaller than $1/r'$ by virtue of Lemma 5. Further, note that if either H_1 is in $\mathbb{K}_c^r$ and $c < r$ or H_1 is in $\bar{\mathbb{K}}_c^r$ and $c > r$, then, in both cases, $H_1^{-1} \in \bar{\mathbb{K}}_{-c'}^{r'}$. Algebraic manipulations reveal that

$$H_1^{-1} \in \bar{\mathbb{K}}_{-c'}^{r'} \quad \Rightarrow \quad H_1^{-1} + c'\,\mathrm{id} \in \bar{\mathbb{K}}_0^{r'} \quad \Rightarrow \quad H_1' \in \mathbb{K}_0^{1/r'} \tag{4.43}$$

whence gain (H_1') is smaller than or equal to $1/r'$. Therefore, the product of gain (H_1') and gain (H_2') is smaller than 1 and we infer that E_u' and E_x' are bounded from Lemma 3. As Lemma 6 states that E_u and E_x are contained in E_u' and E_x', respectively, this concludes the proof. □

Example 23. Let H_1 relate e to

$$x : t \mapsto \frac{\|e(t)\| + 1}{\|e(t)\|} \operatorname{rot}(\alpha)\, e(t), \quad \operatorname{rot}(\alpha) = \begin{bmatrix} \cos(\alpha) & \sin(\alpha) \\ -\sin(\alpha) & \cos(\alpha) \end{bmatrix} \tag{4.44}$$

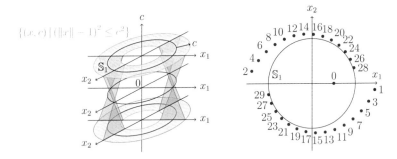

Figure 4.7: The cone $\mathbb{K}_0^1$ (left) and a sequence converging to the circle (right)

which is just the relation H_1 from Example 21 with an additional rotation by some angle α. If one asks whether H_1 is contained in $\mathbb{K}_c^r$ whereby letting S be the circle $\mathbb{S}_1$, one arrives at the inequality $c^2 + 1 \leq r^2$, which is, e.g., satisfied for $c = 0$ and $r = 1$. The cone $\mathbb{K}_0^1$ contains all relations whose members (e, x) satisfy

$$\int_{\mathbb{R}} (\|x(t)\| - 1)^2 \, dt \leq \int_{\mathbb{R}} \|e(t)\|^2 \, dt \tag{4.45}$$

when truncated. Instantaneously, i.e. pointwise in t, this is satisfied by all pairs (x, e) satisfying $(\|x\| - 1)^2 \leq \|e\|^2$, which we next illustrate graphically. For this purpose, parametrize e through $|c| = \|e\|$ for some real c (this parameterization is reasonable as only the norm of e matters). All x and c for which $(\|x\| - 1)^2 \leq c^2$ are contained in the gray areas depicted left in Figure 4.7. Therein, three layers with distinct values of c are plotted, one of which is associated with $c = 0$: on this layer, only the circle $\mathbb{S}_1$ itself satisfies $(\|x\| - 1)^2 \leq c^2$. On all other layers, the inequality is satisfied by annuli, the difference of whose radii is precisely $2c$. Each of these annuli contain the circle and the annuli are joint through cones, four of which are depicted in the figure, two in red and two in blue, with equal colors assigned to opposite (i.e. equally oriented) cones.

Knowing that H_1 is in $\mathbb{K}_0^1$, we can proceed by constructing a controller such as we did in the proof of Theorem 4. To do so, we would have to construct a control relation $-H_2$ contained in $\mathbb{K}_0^{1-\epsilon}$ for some small ϵ. One finds that the control relation $-H_2$ from Example 21 is in this very cone if we chose k to be $1 - \epsilon$. Solving for x and e iteratively whilst letting w be zero yields the iteration

$$x_{i+1} = \left(1 - \epsilon + \frac{1}{\|\pi(x_i)\|}\right) \text{rot}(\alpha) \pi(x_i) = \left(1 - \epsilon + \frac{1}{\|\pi(x_i)\|}\right) \pi(\text{rot}(\alpha) x_i) \tag{4.46}$$

which generates the sequence depicted right in Figure 4.7 for $x_0 = \frac{1}{2}e_1$ and $\epsilon = \alpha = \frac{1}{10}$. The blue circles are enumerated by the iteration which they represent. As expected, the sequence approaches the circle, but jumps back and forth from one side to the other while doing so. The cones with center 0 and radius 1 are comparatively simple. Otherwise, if one seeks more general conic H_2 for S being the circle, it is helpful to depict the condition for $-H_2$ being in $\mathbb{K}_c^r$ by first depicting the inequality defining the

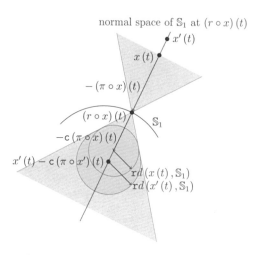

Figure 4.8: The control u must point inside balls centered along normal spaces of $\mathbb{S}_1$

cone pointwise in time, i.e. for some particular argument t of u and x, which would read

$$(u(t) - (\mathsf{c} - \mathbf{r})(\pi \circ x)(t)) \cdot (u(t) - (\mathsf{c} + \mathbf{r})(\pi \circ x)(t)) \leq 0. \tag{4.47}$$

A conic $-H_2$ then can be thought of as containing signals satisfying the above condition in their integral. The condition states that u must point inside a closed ball, the radius of which is $\mathbf{r}d(x(t), \mathbb{S}_1)$, centered at a point on the normal space of $\mathbb{S}_1$ at $r \circ x$, namely at $x(t) - \mathsf{c}(\pi \circ x)(t)$. The union of these balls over all $x(t)$ which instantaneously lie in the normal space of $\mathbb{S}_1$ at $(r \circ x)(t)$ constitutes a cone, which is depicted in gray in Figure 4.8 with two of the contained balls highlighted in gray. The radii are indicated by red arrows and the centers result from scaling $\pi \circ x$ by c, as indicated by the blue arrows and circles. In Figure 4.8 as well as in Figure 4.7 (left), we depicted the cones relevant for $-H_2$ and H_1, respectively. We find that all cones are centered at the circle but the cone constructed in Figure 4.8 could be viewed as a subset of the ambient space of $\mathbb{S}_1$ whilst the cones from Figure 4.7 (left) could only be represented by introducing an additional dimension for u. In both cases, we had to construct a cone for every point of the circle, i.e. a family of cones, smoothly varying with the point of $\mathbb{S}_1$ we consider. This is in contrast to the classical conicity condition of Zames, which is illustrated by a single cone (cf. [97, Fig. 6]), but reduces to his familiar form for S being a singleton.

Passive Relations

Verifying whether the system H_1 is contained in some $\mathbb{K}_\mathsf{c}^\mathbf{r}$ may turn out hard in practice since even instantaneous depiction of the inequality defining $\mathbb{K}_\mathsf{c}^\mathbf{r}$ becomes involved.

However, considering the limiting case

$$\mathsf{c} + \mathsf{r} \to \infty \qquad (4.48)$$

$$\mathsf{c} - \mathsf{r} \to 0 \qquad (4.49)$$

turns out to simplify the interpretation of the relations satisfying the inequality defining $\mathbb{K}_{\mathsf{c}}^{\mathsf{r}}$. We call such relations passive and devote the remainder of this chapter to their analysis. In particular, let

$$\mathbb{P}_{\mathsf{e}}^{\mathsf{s}} = \big\{ H \subset \mathscr{L}_S^{2,n} \times \mathscr{L}_{S'}^{2,n} \big| \text{ for all } (\Phi, \Phi') \text{ in } H, \text{ for all } t \text{ in } \mathbb{R},$$
$$\big\langle \pi \circ \Phi_S^t, \pi' \circ \Phi_{S'}^{\prime t} \big\rangle_{\mathscr{L}^{2,n}} \geq \mathsf{e} \left\| \Phi_S^t \right\|_{\mathscr{L}_S^{2,n}}^2 + \mathsf{s} \left\| \Phi_{S'}^{\prime t} \right\|_{\mathscr{L}_{S'}^{2,n}}^2 \big\} \qquad (4.50)$$

wherein e and s are both real scalars, called the excess and the shortage (or lack) of passivity, respectively. We refer to the members of $\mathbb{P}_{\mathsf{e}}^{\mathsf{s}}$ as being passive. If, further, $\mathsf{e} = 0$ and $\mathsf{s} < 0$, then the relation is said to be output feedback passive with the shortage of passivity $-\mathsf{s}$. If $\mathsf{s} = 0$ and $\mathsf{e} > 0$, then we call a relation contained in $\mathbb{P}_{\mathsf{e}}^{\mathsf{s}}$ input strictly passive with the excess of passivity e. If the feedforward path H_1 is output feedback passive and the feedback path $-H_2$ is input strictly passive with sufficiently large excess of passivity, then E_x is bounded, as our next proposition, which was previously published in [M6, M7], states.

Proposition 7. Let E_x relate w to x subject to (4.8)-(4.10). For all $\mathsf{s} \geq 0$ and $\mathsf{e} > \mathsf{s}$, for all $\mathsf{a} \geq 0$, for all H_1 in $\mathbb{P}_0^{-\mathsf{s}}$, for all $-H_2$ in $\mathbb{P}_{\mathsf{e}}^{\mathsf{a}}$, E_x is bounded.

Proof. Let w, x, e, u, solve the feedback equations. As H_1 is output feedback passive with shortage of passivity s, we have

$$\big\langle w^t, \pi \circ x_S^t \big\rangle_{\mathscr{L}^{2,n}} = \big\langle e^t, \pi \circ x_S^t \big\rangle_{\mathscr{L}^{2,n}} - \big\langle u^t, \pi \circ x_S^t \big\rangle_{\mathscr{L}^{2,n}} \geq -\mathsf{s} \left\| x_S^t \right\|_{\mathscr{L}_S^{2,n}}^2 - \big\langle u^t, \pi \circ x_S^t \big\rangle_{\mathscr{L}^{2,n}} \qquad (4.51)$$

for all t in $\mathbb{R}$. If we further exploit that $-H_2$ is input strictly passive (due to the inclusion $\mathbb{P}_{\mathsf{e}}^{\mathsf{a}} \subset \mathbb{P}_{\mathsf{e}}^0$), we next arrive at

$$\big\langle w^t, \pi \circ x_S^t \big\rangle_{\mathscr{L}^{2,n}} \geq (\mathsf{e} - \mathsf{s}) \left\| x_S^t \right\|_{\mathscr{L}_S^{2,n}}^2 \qquad (4.52)$$

and thus conclude that

$$\left\| w^t \right\|_{\mathscr{L}^{2,n}} \left\| \pi \circ x_S^t \right\|_{\mathscr{L}^{2,n}} = \left\| w^t \right\|_{\mathscr{L}^{2,n}} \left\| x_S^t \right\|_{\mathscr{L}_S^{2,n}} \geq (\mathsf{e} - \mathsf{s}) \left\| x_S^t \right\|_{\mathscr{L}_S^{2,n}}^2 \qquad (4.53)$$

via Cauchy-Schwarz and the identity $\left\| \pi \circ \Phi_S^t \right\|_{\mathscr{L}^{p,n}} = \left\| \Phi_S^t \right\|_{\mathscr{L}_S^{p,n}}$ from Lemma 4. Should e be greater than s, we find that

$$\frac{1}{\mathsf{e} - \mathsf{s}} \left\| w^t \right\|_{\mathscr{L}^{2,n}} \geq \left\| x_S^t \right\|_{\mathscr{L}_S^{2,n}} \qquad (4.54)$$

which reveals that E_x is bounded by taking the limit $t \to \infty$. $\qquad \square$

Example 24 (Dissipativity Characterization of Passivity). Establishing a connection between the input-output point of view, taken in this chapter, and systems modeled by differential equations, such as in the foregoing chapters, is generally not simple. One reason for this is that there is no canonical realization of an input-output relation as differential equation. However, if one was given a realization of H_1 of the form

$$(u, x) \in H_1 \quad :\Leftrightarrow \quad \exists \xi_0 : \quad \xi(0) = \xi_0, \quad \dot{\xi}(t) = f(\xi(t), u(t)) \text{ for almost all } t, \quad (4.55)$$
$$x(t) \in U, \quad x(t) = h(\xi(t), u(t)) \text{ for almost all } t, \quad (4.56)$$

for some $f : \mathbb{R}^p \times \mathbb{R}^n \to \mathbb{R}^p$ and $h : \mathbb{R}^p \times \mathbb{R}^n \to \mathbb{R}^n$, then passivity of H_1 can be characterized via a so-called storage function $V : \mathbb{R}^p \to \mathbb{R}$ through the dissipativity characterization of passivity: let V be nonnegative and let $V \circ \mathsf{P}_1 \circ h^{-1}$, with $\mathsf{P}_1 : (\xi, u) \mapsto \xi$, be zero on S. Assume that, for every ξ_0 which satisfies the above requirements (4.55)-(4.56), there exists $u_{\xi_0} : (-\infty, 0] \to \mathbb{R}^n$ such that $\xi(0) = \xi_0$ and $d(x(t), S) \to 0$ as $t \to -\infty$ (i.e. there exists a point in $(\mathsf{P}_1 \circ h^{-1})(S)$ from which ξ_0 is reachable). If, for all $u' \in \mathbb{R}^n$, and all $\xi' \in \mathbb{R}^p$, the dissipation inequality

$$(\pi \circ h)(\xi', u') \cdot u' \geq \nabla V(\xi') \cdot f(\xi', u') \tag{4.57}$$

holds, then for all $u : \mathbb{R} \to \mathbb{R}^n$ such that $u\big|_{(-\infty, 0]} = u_{\xi_0}$ (in practice, the choice of $u(t)$ for $t \leq 0$ is irrelevant), H_1 is passive. This is verified by the calculation

$$\left\langle u^{t'}, \pi \circ x_S^{t'} \right\rangle_{\mathscr{L}^{2,n}} = \int_{-\infty}^{t'} (\pi \circ x)(t) \cdot u(t)\, \mathrm{d}t \geq V(\xi(t')) - \lim_{t \to \infty} V(\xi(-t)) \geq 0 \tag{4.58}$$

in which we used the chain rule, the dissipation inequality, and that $V \circ \mathsf{P}_1 \circ h^{-1}$ is zero on S. The advantage of the above procedure is that the dissipation inequality does not involve knowledge of solutions of the differential equation realizing H_1; in fact, the dissipation inequality can be verified or falsified solely with knowledge of f and h (plus a candidate for the storage function V).

Given the connection between passivity of a state-space realization and passive input-output relations established above, this example bridges between the present chapter and the passivity-based approach of El-Hawwary and Maggiore to stabilization of closed sets [34], in which the authors generalize the seminal results of Byrnes, Isidori, and Willems [27]. We, similarly, obtain the usual dissipation inequality for passive systems from the dissipation inequality (4.57) if we choose $S = \{0\}$.

Example 25. Let S be the unit circle $\mathbb{S}_1$, embedded in $\mathbb{R}^2$. We wish to construct input strictly passive controllers $-H_2$. To this end, first consider the condition $-\langle \pi \circ x, u \rangle_{\mathscr{L}^2} \geq \mathrm{e} \|x\|^2_{\mathscr{L}^{2,2}_{\mathbb{S}_1}}$, defining input strict passivity of $-H_2$ (omitting the truncation for simplicity), pointwise, i.e. for some particular argument t of x and u. The inequality then necessitates that $t \mapsto u(t)$ makes an acute angle with $t \mapsto -(\pi \circ x)(t)$ at time t. Input strictly passive controllers thus relate x to u so as to let u and $-\pi \circ x$ make an acute angle in their integral. Graphically, this amounts to asking that u points inside the tubular neighborhood $\{x' \in \mathbb{R}^2 | d(x', \mathbb{S}_1) < d(x(t), \mathbb{S}_1)\}$ of $\mathbb{S}_1$, which we depict in Figure 4.9. In the figure, the tubular neighborhood inside which u has to point is colored gray with the distance of $x(t)$ to the circle $\mathbb{S}_1$ indicated by a red arrow. The normal space of $\mathbb{S}_1$ at $(r \circ x)(t)$ is plotted in red, as well and all signals are depicted as blue arrows or circles.

If one now was to actually construct a controller $-H_2$ which is input strictly passive, one could exploit our above observation and measure the angle enclosed between u and

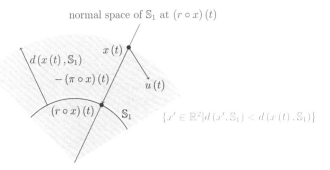

Figure 4.9: The control u makes an acute angle with $-\pi \circ x$

$-\pi \circ x$, which is done through the function

$$k : \mathbb{R} \to \mathbb{R}, \quad t \mapsto \frac{x(t) \cdot u(t)}{\|x(t)\|(1 - \|x(t)\|)} \tag{4.59}$$

by projecting u onto the normal space of $\mathbb{S}_1$ at $r \circ x$ and representing it in the basis E_π. We find that $-H_2$ is input strictly passive with excess of passivity e if and only if

$$\int_\mathbb{R} (\mathsf{e} - k(t)) \, d(x(t), \mathbb{S}_1)^2 \, \mathrm{d} t \leq 0 \tag{4.60}$$

which is true whenever k assumes values greater than or equal to e in its integral. This simplifies the problem of designing H_2 by now merely having to choose a scalar-valued signal k. This construction remains correct if we transition from the unit circle in $\mathbb{R}^2$ to the n-sphere $\mathbb{S}_n$ in $\mathbb{R}^{n+1}$, which is due to the fact that these spheres have codimension 1 in their ambient spaces, whence the projection of u onto their normal spaces can be expressed in terms of a scalar valued function by left-multiplication with $\mathsf{E}_\pi^\top$. In other words, one may choose k independent of the dimension of the sphere which one wishes to stabilize.

The content of this example was previously published in [M6].

Persistence of Excitation

The condition that $-\langle \pi \circ x, u \rangle_{\mathscr{L}^{2,n}} \geq \mathsf{e} \|x\|_{\mathscr{L}^{2,n}_S}^2$, which defines input strict passivity of $-H_2$ can be thought of as requiring that u should be rich enough in the normal spaces of S (which here must not be confused with the fibers resulting from affinely translating the normal spaces to S). More than that, it is even necessary for u to be recurrently rich in the normal spaces of S, which we formalize through the persistence-of-excitation condition known from literature on adaptive identification (among others from [2, 64]). We say that $V : \mathbb{R} \to \mathbb{R}^{n \times p}$ is persistently exciting if there should exist positive real scalars δ and α with the property that

$$\int_{t'}^{t'+\delta} V(t) V(t)^\top \, \mathrm{d} t \geq \alpha I_n \tag{4.61}$$

holds for all t' in $\mathbb{R}$. We focus our attention on controllers H_2 which relate x to

$$t \mapsto -K(t)(\pi \circ x)(t) \tag{4.62}$$

for some function $K : t \mapsto \mathbb{R}^{n \times n}$. This Ansatz appears reasonable as the input strict passivity inequality imposes a condition on the angle enclosed by u and $\pi \circ x$ anyhow (on the other hand, linearity in π was proven to be necessary for optimality in the previous chapter). We then further find that input strict passivity will require $K(t)$ to have positive semidefinite symmetric part at least for some t. To simplify our analysis, we thus assume that the symmetric part of $K(t)$, call it $K_{\mathrm{sym}}(t)$, is positive semidefinite for all t. Let p denote the smallest number among all ranks of $K_{\mathrm{sym}}(t)$, $t \in \mathbb{R}$. Then K_{sym} decomposes as the Gramian of the rows of, say, $\mathsf{G} : \mathbb{R} \to \mathbb{R}^{n \times p}$, i.e. $K_{\mathrm{sym}} = \mathsf{G}\mathsf{G}^\top$. Lastly, define a matrix B to have columns which are an orthonormal basis for a collection of normal spaces of S, i.e. its columns are a least collection of vectors such that their linear hull is contained in the union of the images of all E_π, viz.

$$\mathrm{im}(\mathsf{B}) \subset \bigcup_{x \in U} \mathrm{im}\left(\mathsf{E}_{\pi(x)}\right) = \bigcup_{x \in S} N_x S \tag{4.63}$$

wherein B may be obtained from collecting columns of the matrices $\mathsf{E}_{\pi(x)}$, $x \in U$, and removing linearly dependent columns until the above inclusion is achieved. For instance, consider a (homotopy) cylinder as a submanifold of $\mathbb{R}^3$ and suppose that it has a bulge as depicted left in Figure 4.10, hence letting it have normal vectors that the cylinder would not have without that bulge. More particularly, if one was to collect all normal vectors v with $\|v\| = 1$ for the cylinder (similar to the image of the Gauss map), one would obtain a circle. The collection of these normal vectors for the cylinder with bulge is a subset of the 2-sphere $\mathbb{S}_2$, as depicted right in Figure 4.10 in blue, which contains the circle as a submanifold of $\mathbb{S}_2$. A feasible choice for the image of B would thus be the 2-dimensional subspace of $\mathbb{R}^3$ aligned with the circle, depicted on the right as gray grid. This subspace does not cover the normal vectors caused by the bulge but is the largest subspace satisfying the above inclusion. To see how one transitions from the left-hand side of Figure 4.10 to the right-hand side, we selected a point x from $\mathbb{R}^3$, indicated by a red circle on the left, computed the normal vector $\pi(x)$ and, assuming that its length is 1, plotted it as a red arrow on the left as well as on the right.

Having this construction for B at hand and assuming further that the domain of H_2 contains $\mathscr{L}_S^{2,n}$, then, in the described setting, we find that persistence of excitation of $\mathsf{B}^\top\mathsf{G}$ is necessary for input strict passivity of $-H_2$, which was previously published in [M6].

Proposition 8. Let the columns of B be an orthonormal basis for a collection of normal spaces of S. Let H_2 relate x to (4.62) and let K decompose as the Gramian of the rows of G. If $-H_2$ is input strictly passive, then $t \mapsto \mathsf{B}^\top\mathsf{G}(t)$ is persistently exciting.

Proof. Consider some x' from U such that $\pi(x')$ is in the image of B and some t' from $\mathbb{R}$. Choose $\delta > 0$. Define

$$x : t \mapsto \begin{cases} x' & \text{for } t \in [t', t' + \delta] \\ r(x') & \text{elsewhere} \end{cases} \tag{4.64}$$

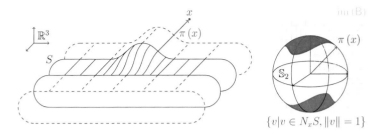

Figure 4.10: A submanifold (left) and all its normal vectors (right), containing $\mathrm{im}\,(\mathsf{B})$

and find that this x should lie in the domain of H_2 since we assumed that all members of $\mathscr{L}_S^{2,n}$ are contained in its domain whilst x is in $\mathscr{L}_S^{2,n}$ for assuming only finitely many values and satisfying $\|x\|_{\mathscr{L}_S^{2,n}} = \delta\pi\,(x') \cdot \pi\,(x') < \infty$. Substituting this last expression and (4.62) into the inequality defining input strict passivity of $-H_2$ with excess of passivity e, we arrive at

$$\pi\,(x') \cdot \int_{t'}^{t'+\delta} \mathsf{G}\,(t)\,\mathsf{G}\,(t)^\top \,\mathrm{d}\,t\,\pi\,(x') \geq \mathsf{e}\delta\pi\,(x') \cdot \pi\,(x')\,. \tag{4.65}$$

As we chose x' such that $\pi\,(x')$ lies in the image of B, we can find a vector v such that $\pi\,(x') = \mathsf{B}v$. Substituting this expression for $\pi\,(x')$ and exploiting the orthonormality of B, we are left with

$$v \cdot \int_{t'}^{t'+\delta} \mathsf{B}^\top\mathsf{G}\,(t)\,\mathsf{G}\,(t)^\top \mathsf{B}\,\mathrm{d}\,t\,v \geq \mathsf{e}\delta\mathsf{B}v \cdot \mathsf{B}v = \mathsf{e}\delta v \cdot v\,. \tag{4.66}$$

This inequality must remain satisfied for any x' in U such that $\pi\,(x')$ lies in the image of B. Therefore, the above inequality must remain true for all v, which hence characterizes $t \mapsto \mathsf{B}^\top\mathsf{G}\,(t)$ as being persistently exciting. $\qquad\square$

Choosing B such that its image is a largest subspace satisfying (4.63) (we did not impose other assumptions on B than (4.63) and thus could have also formulated the claim of the foregoing proposition for all orthonormal B satisfying (4.63)), the condition that $\mathsf{B}^\top\mathsf{G}$ be persistently exciting can be interpreted as G being recurrently rich in the normal spaces of S. It should be noted that the requirement on the richness of G therein scales with the image of the Gauss map, i.e.: should the Gauss map be surjective, which is the case when S is compact and orientable, then the normal spaces of S jointly span the ambient space, letting the identity matrix be a feasible choice for B. But as the image of the Gauss map becomes smaller, the number of columns of B reduces accordingly and hence G is required to be recurrently rich in fewer subspaces.

Example 26 (Synchronization). Let S be the diagonal $\mathscr{D}$ of $\mathbb{R}^n$ (a product space could also be taken into account). We wish to construct an input strictly passive controller which brings x to $\mathscr{D}$. As an Ansatz, let H_2 relate x to

$$u : t \mapsto -K\,(t)\,x\,(t)\,. \tag{4.67}$$

With this Ansatz, $-H_2$ is input strictly passive with excess of passivity e when

$$\int_{\mathbb{R}} K(t) x(t) \cdot \mathsf{P}_\pi x(t) \, \mathrm{d}\, t \geq \mathsf{e} \int_{\mathbb{R}} \|\mathsf{P}_\pi x(t)\|^2 \, \mathrm{d}\, t \qquad (4.68)$$

remains satisfied, wherein P_π is defined as in the foregoing chapter (and more particularly as in Example 15). If we choose K constant, i.e. $K : t \mapsto K'$ and insist that the above inequality should be satisfied pointwise in t, i.e.

$$\exists \mathsf{e} > 0 : \quad \forall t \in \mathbb{R}, \quad K'x(t) \cdot \mathsf{P}_\pi x(t) \geq \mathsf{e} \|\mathsf{P}_\pi x(t)\|^2, \qquad (4.69)$$

then, denoting the symmetric part of K' by K'_{sym}, and letting E_π be the matrix consisting of columns which form an orthonormal basis of the orthogonal complement of $\mathcal{D}$ in $\mathbb{R}^n$, we find that this condition is equivalent to the following three statements: (i) $\mathsf{E}_\pi^\top K'_{\mathsf{sym}} \mathsf{E}_\pi$ is positive definite (i.e. the quadratic form $x \mapsto x \cdot K'x$ assumes positive values for nonzero arguments from the orthogonal complement of $\mathcal{D}$ in $\mathbb{R}^n$), (ii) the nullspace of K' is contained in $\mathcal{D}$, (iii) the image of $\mathcal{D}$ under $x \mapsto K'x$ is contained in $\mathcal{D}$ (i.e. $\mathcal{D}$ is K-invariant). One notices that the properties (i)-(iii) are met by Laplacian matrices of undirected, connected, (possibly weighted) graphs. Moreover, should K' be the Laplacian matrix of such a graph, then the excess of passivity e is precisely the algebraic connectivity of the graph. This being said, if we were not to assume that K is constant but now let $K : t \mapsto K(t)$, and invoke the previous proposition, then we find that the condition that $-H_2$ is input strictly passive necessitates persistence of excitation of $\mathsf{E}_\pi^\top \mathsf{G}$, where $K_{\mathsf{sym}}(t)$, the symmetric part of $K(t)$, decomposes as the Gramian of the rows of $\mathsf{G}(t)$. If we again let $K(t)$ be the Laplacian matrix of an undirected, connected, (possibly weighted), graph, varying with t, then $\mathsf{G}^\top$ is just the incidence matrix of the given graph: the reason for this is that the Laplacian matrix decomposes as the Gramian of the columns of the incidence matrix.

This latter result, that the transpose of the incidence matrix should be persistently exciting in the orthonormal basis of the orthogonal complement of $\mathcal{D}$, was independently discovered in [4, equation (70)].

Similar to the approach taken in this example, Scardovi et al. [80] applied input-output techniques to synchronization problems with application to biochemical reaction networks.

The content of this example was previously published in [M6].

Zames-Falb Multipliers

The motivation to construct controllers from input-output data of H_1 partially stems from experimental setups or similar applications in which no other data than input-output data may be gathered. With this in mind, the key idea behind Proposition 7, conceptually, is that it is sufficient to figure whether H_1 is (output feedback) passive from the given data. Should it turn out hard to verify whether H_1 is passive from given data, then applying a filter Z to the input e of H_1 and doctoring Z until the relation which relates $Z(e)$ to x looks passive in the sense that it is contained in

$$\bar{\mathbb{P}}_\mathsf{e} = \left\{ H \subset \mathscr{L}_S^{2,n} \times \mathscr{L}_{S'}^{2,n} \,\middle|\, \text{for all } (\Phi, \Phi') \text{ in } H, \quad \langle \pi \circ \Phi_S, \pi' \circ \Phi'_{S'} \rangle_{\mathscr{L}^{2,n}} \geq \mathsf{e} \|\Phi_S\|_{\mathscr{L}_S^{2,n}}^2 \right\}$$
$$(4.70)$$

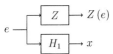

Figure 4.11: Parallel interconnection of plant H_1 and multiplier Z

significantly simplifies application of Proposition 7 (the restriction to $\mathscr{L}_S^{2,n}$ instead of the extension $\bar{\mathscr{L}}_S^{2,n}$ is meant to simplify presentation – we do not consider a shortage of passivity in this part of the chapter for the same reason). In fact, this is the key idea behind the Zames-Falb multipliers [99]. A great advantage in introducing the multiplier Z (these filters are referred to as multipliers in this context) is that it may be adjusted offline even after the input-output data of H_1 has been obtained, say experimentally, without the need of performing new experiments. This way of connecting Z and H_1 in parallel is depicted in Figure 4.11. The Zames-Falb multipliers are applicable in connection with the feedback theorem for passive systems (cf. [33, chapter VI, section 9]) and thus apply to the case where S is the origin of $\mathbb{R}^n$. In the remainder, we develop a framework for multipliers which are applicable to rather general submanifold stabilization problems.

To this end, let $Z : \mathscr{L}^{2,n} \to \mathscr{L}^{2,n}$ be a linear bijection which admits the factorization $Z = Z_1 \circ Z_2$ such that $Z_2 : \mathscr{L}^{2,n} \to \mathscr{L}^{2,n}$ is a causal linear bijection and $Z_1 : \mathscr{L}^{2,n} \to \mathscr{L}^{2,n}$ has a causal bijective adjoint Z_1^*. It shall be noted that Z itself is not required to be causal, which is the reason why the relation which relates $Z(e)$ to x might not be passive even though it is contained in $\bar{\mathbb{P}}_0$ for an appropriate choice of Z. Indeed, should a relation be causal, then its membership in $\bar{\mathbb{P}}_0$ implies passivity but this implication is false for noncausal relations (unfortunately, many relevant multipliers are not causal). The reason for requiring the factorization $Z_1 \circ Z_2$ of Z is to study the feedback interconnections depicted in Figures 4.12 and 4.13 instead of the original feedback interconnection from Figure 4.1. Therein, we realize that Z_1^* can not be the usual adjoint as its domain must be $\mathscr{L}_S^{2,n}$ for connecting it to the output of H_1. In fact, $Z_1^* : \mathscr{L}_S^{2,n} \to \pi \circ \left(\mathscr{L}_S^{2,n} \right)$ must be defined to satisfy

$$\langle \pi \circ \Phi, Z_1(\Phi') \rangle_{\mathscr{L}^{2,n}} = \langle Z_1^*(\Phi), \Phi' \rangle_{\mathscr{L}^{2,n}} \tag{4.71}$$

such that we recover the usual adjoint if S was $\{0\}$. We require Z_1^* to be a bijection and that itself and its inverse have finite gain (in the same sense as we defined gains of relations). Given this setup, our following lemma reasons why we may investigate the feedback interconnections depicted in Figures 4.12 and 4.13 instead of our original feedback interconnection. In particular, let E_x' relate w to x' subject to

$$
\begin{aligned}
z_1 &= z + z_2 & z &= Z(w) & (4.72) \\
(e', x') &\in H_1 & e' &= Z^{-1}(z_1) & (4.73) \\
(x', u') &\in H_2 & z_2 &= Z(u') & (4.74)
\end{aligned}
$$

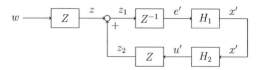

Figure 4.12: Feedback interconnection of system H_1, controller H_2, and multiplier Z

and let E_x'' relate w to x'' subject to

$$z' = Z_2(w) \qquad\qquad e'' = Z_2^{-1}(z_1') \qquad (4.75)$$
$$z_1' = z' + z_2' \qquad\qquad z^* = Z_1^*(x'') \qquad (4.76)$$
$$(e', x') \in H_1 \qquad\qquad x'' = Z_1^{*-1}(z^*) \qquad (4.77)$$
$$(x', u') \in H_2 \qquad\qquad z_2' = Z_2(u'') \qquad (4.78)$$

and find that boundedness of E_x, E_x', and E_x'' are equivalent, given the above circumstances. This was previously published in [M6].

Lemma 7. Let E_x, E_x', E_x'' relate w to x, x', x'' subject to (4.8)-(4.10), (4.72)-(4.74), (4.75)-(4.78), respectively. Let $Z = Z_1 \circ Z_2 : \mathscr{L}^{2,n} \to \mathscr{L}^{2,n}$ be a linear bijection such that $Z_2 : \mathscr{L}^{2,n} \to \mathscr{L}^{2,n}$ is a causal linear bijection and $Z_1 : \mathscr{L}^{2,n} \to \mathscr{L}^{2,n}$ has a causal bijective adjoint Z_1^*, defined by (4.71). Then the following three statements are equivalent: (i) E_x is bounded, (ii) E_x' is bounded, (iii) E_x'' is bounded.

Proof. We first prove that (i) is equivalent to (ii): let z_1, w, u', e', x' solve the feedback equations (4.72)-(4.74). By linearity of Z, we have that $z_1 = Z(w + u')$. Substituting the right-hand equation in (4.73), we recover $e' = w + u'$. Hence, w, u', e', x' also solve the feedback equations (4.8)-(4.10). Since Z is bijective, every solution of (4.8)-(4.10) also solves (4.72)-(4.74). Thus, (i) is equivalent to (ii). We next show that (ii) is equivalent to (iii): let z_1', w, u'', e'', x'' solve the feedback equations (4.75)-(4.78). By linearity of Z_2, z_1' equals $Z_2(w + u'')$. Substituting the right-hand equation in (4.75), we recover $e'' = w + u''$. Thus, w, u'', e'', x'' also solve (4.8)-(4.10). Since Z_1^* and Z_2 are both bijections, every solution of (4.8)-(4.10) also solves (4.75)-(4.78), revealing that (i) is equivalent to (iii). This establishes the claim. □

The basic reasoning now is as follows: define by H_1' the relation which relates z_1 to x' and by H_2' the relation which relates x' to z_2, both subject to (4.72)-(4.74). Further, define by H_1'' the relation which relates z_1' to z^* and by H_2'' the relation which relates z^* to z_2', both subject to (4.75)-(4.78). Calling a member of $\bar{\mathbb{P}}_e$ positive when $e = 0$ and strictly positive when $e > 0$, one finds that positivity of H_1' together with strict positivity of $-H_2'$ implies positivity of H_1'' and strict positivity of $-H_2''$. This was previously published in [M6].

Lemma 8. Let H_1', H_2' relate z_1 to x', x' to z_2, respectively, subject to (4.72)-(4.74) and let H_1'', H_2'' relate z_1' to z^*, z^* to z_2', respectively, subject to (4.75)-(4.78). Let $Z = Z_1 \circ Z_2 : \mathscr{L}^{2,n} \to \mathscr{L}^{2,n}$ be such that $Z_2 : \mathscr{L}^{2,n} \to \mathscr{L}^{2,n}$ is a causal bijection and $Z_1 : \mathscr{L}^{2,n} \to \mathscr{L}^{2,n}$ has a causal bijective adjoint Z_1^*, defined by (4.71), whose inverse has finite gain. Then, for all $e > 0$, there exists $e' > 0$ such that for all H_1' in $\bar{\mathbb{P}}_0$ and $-H_2'$ in $\bar{\mathbb{P}}_e$, H_1'' is in $\bar{\mathbb{P}}_0$ and $-H_2''$ is in $\bar{\mathbb{P}}_{e'}$.

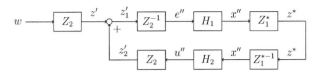

Figure 4.13: Feedback interconnection with factorized multiplier

Proof. Let H_2' relate x' to z_2. As $-H_2'$ is strictly positive, we find that

$$- \langle \pi \circ x', Z\,(u') \rangle_{\mathscr{L}^{2,n}} = - \langle \pi \circ x', (Z_1 \circ Z_2)\,(u') \rangle_{\mathscr{L}^{2,n}} \tag{4.79}$$

$$= - \langle Z_1^*\,(x'), Z_2\,(u') \rangle_{\mathscr{L}^{2,n}} \geq \mathsf{e}\,\|x'\|_{\mathscr{L}_S^{2,n}}^2 \tag{4.80}$$

from the feedback equations (4.72)-(4.74) and (4.75)-(4.78). As we expected that Z_1^{*-1} has finite gain, the positive scalar

$$\mathsf{e}' = \mathsf{e}\,\mathrm{gain}\left(Z_1^{*-1}\right)^2 \tag{4.81}$$

establishes the overestimate

$$- \langle Z_1^*\,(x'), Z_2\,(u') \rangle_{\mathscr{L}^{2,n}} \geq \mathsf{e}'\,\|z^*\|_{\mathscr{L}^{2,n}}^2 \tag{4.82}$$

and thus, by bijectivity of Z_1^* and Z_2, proves strict positivity of $-H_2''$. To find that positivity of H_1' causes positivity of H_1'', replace $-u'$ by e' and e by 0. $\square$

The previous lemma brings us into the position to state our main result for Zames-Falb multipliers. The workflow is as follows: (i) verify positivity of H_1' by adjusting the filter Z, (ii) choose a controller H_2 such that $-H_2'$ is strictly positive, (iii) conclude boundedness of E_x from the previous two lemmata. This procedure is summarized in the following proposition, which was previously published in [M6].

Proposition 9. Let E_x relate w to x subject to (4.8)-(4.10) and let H_1', H_2' relate z_1 to x', x' to z_2, respectively, subject to (4.72)-(4.74). Let $Z = Z_1 \circ Z_2 : \mathscr{L}^{2,n} \to \mathscr{L}^{2,n}$ be a linear bijection such that $Z_2 : \mathscr{L}^{2,n} \to \mathscr{L}^{2,n}$ is a causal linear bijection and $Z_1 : \mathscr{L}^{2,n} \to \mathscr{L}^{2,n}$ has a causal bijective adjoint Z_1^*, defined by (4.71), whose inverse has finite gain. For all $\mathsf{e} > 0$, for all H_1' in $\bar{\mathbb{P}}_0$, for all $-H_2'$ in $\bar{\mathbb{P}}_{\mathsf{e}}$, E_x is bounded.

Proof. From our previous lemma, we conclude that H_1'' is positive and $-H_2''$ is strictly positive. By causality of Z_1^* and Z_2, H_1'' and $-H_2''$ are causal, as well. Thus, H_1'' is passive and $-H_2''$ is input strictly passive, whence we conclude boundedness of E_x'' from Proposition 7. Application of Lemma 7 reveals boundedness of E_x and thus concludes the proof. $\square$

We remark that the proposition does not presume passivity of H_1' nor input strict passivity of $-H_2'$. Further, we note that the factorization of Z by $Z_1 \circ Z_2$ was expected to meet the same requirements that are stated in the usual Zames-Falb multipliers. This allows us to use Zames-Falb multipliers from literature whose factorization was thoroughly studied therein, cf. [33, chapter VI, subsections 9.4 and 9.5], such as the convolution multipliers initially proposed by Zames and Falb [99]. Yet, constructing controllers H_2 for which we a priori know that $-H_2'$ will be strictly positive for a given class of multipliers, such as for monotone controllers and convolution multipliers if S is a singleton, cf. [99], turns out to be a hard task for general submanifolds. Similarly, relaxing the causality assumptions on the factorization, as it can be done via integral quadratic constraints if S is a singleton, cf. [65], causes technical difficulties for general submanifolds. An intriguing observation is that Safonov and Kulkarni [76] revealed that gradient controllers are necessary to satisfy the monotonicity assumptions of the Zames-Falb framework whilst we had employed gradient controllers in chapter 2 in order to asymptotically stabilize the given submanifold.

One is tempted to think that the classical input-output approaches to setpoint stabilization are applicable to submanifold stabilization problems by replacing H_1 by the relation which relates e to $\pi \circ x$. Yet, this approach is flawed since π is generally not injective.

Summary

In comparison to the two foregoing chapters, in which we studied submanifold stabilization problems for systems modeled by differential equations, we turned our attention to systems solely described by their input-output behavior in the present chapter. We developed a basic framework with which we were able to distinguish between desirable and undesirable outputs, namely via the collection $\mathscr{L}_S^{p,n}$ of all measurable signals with bounded integral p-fold distance to the submanifold S. Although these sets are no vectors spaces, we could formulate a small-gain theorem in our setting. Thereafter, we showed that composition with π brings members of $\mathscr{L}_S^{p,n}$ to the familiar $\mathscr{L}^{p,n}$-spaces, hence endowing us with the capability of treating $\mathscr{L}_S^{p,n}$ as if they were Banach spaces (or Hilbert spaces for $p = 2$). Knowing how to transition to $\mathscr{L}^{p,n}$, we could define conic and passive relations for the case that $p = 2$ and prove a feedback theorem for conic relations as well as a feedback theorem for passive relations, respectively. The feedback theorem for conic relations was based upon a transformation of the feedback interconnection and the fact that cones with center zero have bounded gain. The feedback theorem for passive systems would assume that the feedforward path is output feedback passive and then show that submanifold stabilization is achieved under the condition that the feedback path is input strictly passive with a sufficiently large excess of passivity. For passive relations, we provided an interpretation as a persistence-of-excitation condition, which turned out necessary for input strict passivity. To simplify verifying whether a plant is passive from input-output data, we proposed a multiplier technique similar to the Zames-Falb multipliers. Our multipliers required the same type of factorization as the usual Zames-Falb multipliers. We presented multiple illustrations of our framework.

Discussion

One of the first approaches to controller design via general input-output techniques was pursued by Zames [97, 98], wherein the small-gain theorem, the feedback theorem for conic relations, and the feedback theorem for passive relations were stated for setpoint regulation. Although we stated these results for submanifold stabilization problems, we recover the aforementioned classical results if we choose S to be the origin. Similar to the results of Zames, our feedback theorems provide open sets of controllers and plants for which the control problem is solved. This is advantageous in robust control problems, in which the system (or the controller) is subject to uncertainties.

5 Conclusion

We studied control problems in which a quantity of interest, x, is asked to approach a given submanifold S of the space in which it evolves. In our approaches to these problems, we took three points of view: firstly, in chapters 2 and 3, we let x be a solution to a differential equation modeling the system to be controlled. Therein, chapter 2 was devoted to rendering S an asymptotically stable invariant set by finding control vector fields under which the controlled system assumes the form of a gradient system with drift. In chapter 3, we asked to bring x towards S in an optimal fashion, wherein optimality was quantified by a functional whose value increases as control energy or integral distance of x to S increase. Lastly, in chapter 4, we merely requested that control energy and integral distance of x to S, respectively, remain finite. In doing so, the model for our control system shifted from differential equations to input-output relations. In all three chapters, we arrived at constructive results in the sense that the characterization of the proposed controls had structural and quite graphical, if not intuitive, interpretations. Throughout the thesis, we stumbled upon various illustrative examples, some of which are relevant on their own.

This chapter concludes the thesis. Below, we summarize the results derived in the previous chapters in greater detail, discuss general aspects (i.e. aspects which are common to all chapters), and point to a few further questions which lie beyond the scope of the present thesis.

5.1 Summary

In this section, we recall and discuss the main results of the thesis.

Chapter 2 was devoted to asymptotic stabilization of invariant submanifolds of differential equations via gradient vector fields. Therein, we considered submanifolds embedded in $\mathbb{R}^n$ in section 2.1 and submanifolds embedded in an invariant Riemannian manifold in section 2.2. The key concept of the chapter is reflected by Theorem 1, which states that one-sided Lipschitz continuous drift vector fields can be compensated by gradient vector fields modulo scaling, whenever the scalar field defining the gradient vector field is strongly convex when restricted to the (affinely translated) normal spaces of the given submanifold. Moreover, Proposition 1 would allow us to increase the rate at which solutions approach the invariant submanifold ad libitum. If the submanifold is no invariant set, then, under milder assumptions, it may still be rendered practically stable in the sense of Proposition 2. Theorem 2 in section 2.2 extends the findings of section 2.1 to state spaces which are Riemannian manifolds. Thereafter, section 2.2 is particularly devoted to the question under which circumstances it is possible to choose controls so as to let the closed loop attain the form of a gradient system with drift. We offer an algebraic characterization of the affirmative and negative answers to this question in Proposition 3. Should the invariant submanifold whose asymptotic stabilization we ask for be an equivalence class, then the aforementioned algebraic characterization can be recast in terms of horizontal and vertical spaces, as claimed in Proposition 4.

The controls proposed in chapter 2 would usually involve large control energy. To resolve this issue, we asked for optimal solutions to submanifold stabilization problems within chapter 3, optimality being quantified via a functional whose value increases as control energy or integral distance of the solution of the considered differential equation to the desired submanifold increase. Our main result of the chapter was that such optimal controls are state feedbacks which linearly depend on a function mapping to the normal spaces of the submanifold, whilst the matrix defining the linear relationship has the tangent spaces of the submanifold contained in its nullspace: necessity of such structured controls for optimality is stated in Theorem 3. Thereafter, Proposition 5 claims that special structured controls are also sufficient for optimality under the condition that the desired submanifold is an invariant set. Should the components of all involved vector fields which lie in the normal spaces of the submanifold be constant, then the optimal control can be expressed via the unique positive definite solution of a certain algebraic Riccati equation, which is summarized in Proposition 6.

Instead of asking for controls with minimal control energy and minimal integral distance of the quantity of interest, x, to the submanifold, we merely asked for finite control energy and finite (integral) distance of x to the submanifold in chapter 4. This allowed us to consider systems which relate an input to x in a rather general fashion instead of systems modeled by differential equations such as in chapters 2 and 3. Our main result of this chapter, Theorem 4, states that, should the system relate inputs to outputs contained in a family of certain cones, then controllers which relate x to the complementary cones let the integral distance of x to the desired submanifold remain finite. Letting the radii of these cones tend to infinity, we recover a generalization of the feedback theorem for passive systems, Proposition 7. In this context, input strict passivity can be interpreted as persistence of excitation of the control signal in the normal spaces of the submanifold, which is stated in Proposition 8. In order to simplify verifying whether a plant is passive, we propose a multiplier technique similar to the Zames-Falb multipliers in Proposition 9.

5.2 Discussion

All our results required that the considered submanifold is an embedded submanifold. More particular, we had to work within tubular neighborhoods of the submanifold, existence of which is only guaranteed for embedded submanifolds. The reason for the restriction to tubular neighborhoods stems from the maps π and r, the former of which returns the fiber component of its argument and the latter of which is the retraction from the tubular neighborhood onto the submanifold. Both maps were essential in all results but π was most important: in chapter 2, π defined a Lyapunov function, in chapter 3, we found that optimal controls depend linearly on π and in chapter 4, π allowed us to treat functions with finite integral p-fold distance to the submanifolds as if they were members of $\mathscr{L}^{p,n}$.

In chapter 2, we had to assume that the given submanifold is compact for two reasons:

firstly, Lyapunov's direct method for closed, noncompact sets involves subtle technical difficulties (cf. [8, chapter V, section 4]). Secondly, we had implicitly required that a preimage of $x \mapsto \|\pi(x)\|^2$ is a tubular neighborhood, which is guaranteed only for compact embedded submanifolds (following the construction from [16, chapter II, section 11]). As we did not explicitly ask for asymptotic stability explicitly in chapters 3 and 4, we could overcome the compactness assumption in these chapters.

The restriction to tubular neighborhoods appears inherent, although the Riemannian metric in section 2.2 would allow us to bend the boundary of the tubular neighborhood through rotation of the normal spaces.

5.3 Outlook

Although we offered constructive solutions in all chapters and hinted towards computational solutions to some aspects of the proposed controller design techniques, we owed explicit computational methods for controller design. In particular, the following constructions from the thesis could explicitly be computed in appropriate algorithms:

The construction of φ suggested by formula (2.16) in section 2.1 (formula (2.58) in section 2.2, respectively) could be accomplished iteratively. Possibly, this could be efficiently combined with computation of an ϵ-net of the submanifold and with explicit computation of the gradient of ϕ.

In section 2.2, we found that the coefficients (2.70) could vanish for some Riemannian metric whilst being nonvanishing for another Riemannian metric. Explicit computation of an appropriate metric tensor for given control vector fields and given submanifold appears feasible, possibly through the S-procedure or families of linear matrix inequalities.

The Riccati differential equation (3.19) arising in chapter 3 does not allow for the usual closed form solution due to the dependency of its coefficients on the solution of the controlled differential equation itself. However, an iterative procedure could potentially allow to solve (3.19) backwards approximately by assuming that the solution of the controlled differential equation is a perturbation of the solution of the uncontrolled differential equation to then iteratively improve the estimate of the solution of the controlled differential equation via an intermediate approximate solution of the Riccati differential equation (3.19).

A modern motivation for the input-output considerations of chapter 4 stems from large numerical models for which state space models are often not available. In such scenarios, one will not know all input-output pairs that the system may produce. Yet, a (large) finite collection of input-output pairs can be produced in order to approximate the parameters arising in Lemma 3, Theorem 4, or Proposition 7. Figuring these parameters with appropriate precision so as to guarantee submanifold stabilization from large amounts of input-output data appears feasible via methods from numerical analysis.

Bibliography

[1] H. Amann. *Gewöhnliche Differentialgleichungen*. de Gruyter, 1995.

[2] B. D. O. Anderson. An approach to multivariable system identification. *Automatica*, 13:401–408, 1977.

[3] M. Aoki. Control of large-scale dynamic systems by aggregation. *IEEE Transactions on Automatic Control*, 13:246–253, 1968.

[4] M. Arcak. Passivity as a design tool for group coordination. *IEEE Transactions on Automatic Control*, 52:1380–1390, 2007.

[5] K. B. Ariyur and M. Krstić. *Real-Time Optimization by Extremum-Seeking Control*. Wiley, 2003.

[6] S. Azuma, R. Yoshimura, and T. Sugie. Broadcast control of multi-agent systems. *Automatica*, 49:2307–2316, 2013.

[7] A. Banaszuk and J. Hauser. Feedback linearization of transverse dynamics for periodic orbits. *Systems & Control Letters*, 26:95–105, 1995.

[8] N. P. Bhatia and G. P. Szegő. *Stability Theory of Dynamical Systems*. Springer, 1970.

[9] N. P. Bhatia and G. P Szegő. *Dynamical Systems: Stability Theory and Applications*. Springer, 1967.

[10] S. P. Bhattacharyya, J. B. Pearson, and W. M. Wonham. On zeroing the output of a linear system. *Information and Control*, 20:135–142, 1972.

[11] G. Birkhoff and G.-C. Rota. *Ordinary Differential Equations*. Wiley, 1978.

[12] A. M. Bloch. *Nonholonomic Mechanics and Control*. Springer, 2003.

[13] A. M. Bloch, M. Reyhanoglu, and N. H. McClamroch. Control and stabilization of nonholonomic dynamic systems. *IEEE Transactions on Automatic Control*, 37:1746–1757, 1992.

[14] F. Borrelli and T. Keviczky. Distributed LQR design for identical dynamically decoupled systems. *IEEE Transactions on Automatic Control*, 53:1901–1912, 2008.

[15] F. Brauer. Perturbations of nonlinear systems of differential equations. *Journal of Mathematical Analysis and Applications*, 14:198–206, 1966.

[16] G. E. Bredon. *Topology and Geometry*. Springer, 1993.

[17] R. W. Brockett. System theory on group manifolds and coset spaces. *SIAM Journal on Control*, 10:265–284, 1972.

[18] R. W. Brockett. Nonlinear systems and differential geometry. *Proceedings of the IEEE*, 64:61–72, 1976.

[19] R. W. Brockett. Asymptotic stability and feedback stabilization. In *Differential Geometric Control Theory*, pages 181–191. Birkhäuser, 1983.

[20] R. W. Brockett. Pattern generation and the control of nonlinear systems. *IEEE Transactions on Automatic Control*, 48:1699–1711, 2003.

[21] R. W. Brockett. Optimal control of the Liouville equation. *AMS/IP Studies in Advanced Mathematics*, 39:23–35, 2007.

[22] R. W. Brockett. On the control of a flock by a leader. *Proceedings of the Steklov Institute of Mathematics*, 268:49–57, 2010.

[23] R. W. Brockett. Notes on the control of the Liouville equation. In *Control of Partial Differential Equations*, volume 2048 of *Lecture Notes in Mathematics*, chapter 2, pages 101–129. Springer, 2012.

[24] R. W. Brockett. Synchronization without periodicity. In K. Hüper and J. Trumpf, editors, *Mathematical System Theory - Festschrift in Honor of Uwe Helmke on the Occasion of his Sixtieth Birthday*, pages 65–74. CreateSpace, 2013.

[25] F. Bullo, J. Cortés, and S. Martínez. *Distributed Control of Robotic Networks*. Princeton University Press, 2009.

[26] C. I. Byrnes and A. Isidori. On the attitude stabilization of rigid spacecraft. *Automatica*, 27:87–95, 1991.

[27] C. I. Byrnes, A. Isidori, and J. C. Willems. Passivity, feedback equivalence, and the global stabilization of minimum phase nonlinear systems. *IEEE Transactions on Automatic Control*, 36:1228–1240, 1991.

[28] F. M. Callier and J. L. Willems. Criterion for the convergence of the solutions of the Riccati differential equation. *IEEE Transactions on Automatic Control*, 26:1232–1242, 1981.

[29] E. J. Candès, M. B. Wakin, and S. P. Boyd. Enhancing sparsity by reweighted ℓ_1 minimization. *Journal of Fourier Analysis and Applications*, 14:877–905, 2008.

[30] Y. Cao and W. Ren. Optimal linear-consensus algorithms: An LQR perspective. *IEEE Transactions on Systems, Man, and Cybernetics, Part B: Cybernetics*, 40:819–830, 2009.

[31] J. Cheeger. Critical points of distance functions and applications to geometry. In *Geometric Topology: Recent Developments*, volume 1504 of *Lecture Notes in Mathematics*, pages 1–38. Springer, 1991.

[32] G. De Nicolao and M. Gevers. Difference and differential Riccati equations: A note on the convergence to the strong solution. *IEEE Transactions on Automatic Control*, 37:1055–1057, 1992.

[33] C. A. Desoer and M. Vidyasagar. *Feedback Systems: Input-Output Properties*. Academic Press, 1975.

[34] M. I. El-Hawwary and M. Maggiore. Reduction principles and the stabilization of closed sets for passive systems. *IEEE Transactions on Automatic Control*, 55:982–987, 2010.

[35] M. I. El-Hawwary and M. Maggiore. Reduction theorems for stability of closed sets with application to backstepping control design. *Automatica*, 49:214–222, 2013.

[36] M. Fardad, F. Lin, and M. R. Jovanović. Design of optimal sparse interconnection graphs for synchronization of oscillator networks. *IEEE Transactions on Automatic Control*, 59:2457–2462, 2014.

[37] J. A. Fax and R. M. Murray. Information flow and cooperative control of vehicle formations. *IEEE Transactions on Automatic Control*, 49:1465–1476, 2004.

[38] G. Floquet. Sur les équations différentielles linéaires à coefficients périodiques. *Annales scientifiques de l'École normale supérieure*, 12:47–88, 1883.

[39] F. Forni and R. Sepulchre. A differential Lyapunov framework for contraction analysis. *IEEE Transactions on Automatic Control*, 59:614–628, 2014.

[40] F. Forni and R. Sepulchre. Differentially positive systems. *IEEE Transactions on Automatic Control*, 61:346–359, 2016.

[41] B. A. Francis. The linear multivariable regulator problem. *SIAM Journal on Control and Optimization*, 15:486–505, 1977.

[42] R. A. Freeman, M. Krstić, and P. V. Kokotović. Robustness of adaptive nonlinear control to bounded uncertainties. *Automatica*, 34:1227–1230, 1998.

[43] P. Fuhrmann. On controllability and observability of systems connected in parallel. *IEEE Transactions on Circuits and Systems*, 22:57, 1975.

[44] J. Guckenheimer and P. Holmes. *Nonlinear Oscillations, Dynamical Systems and Bifurcations of Vector Fields*. Springer, 1983.

[45] W. Hahn. *Stability of Motion*. Springer, 1967.

[46] J. K. Hale. Diffusive coupling, dissipation, and synchronization. *Journal of Dynamics and Differential Equations*, 9:1–52, 1997.

[47] R. Hermann and A. Krener. Nonlinear controllability and observability. *IEEE Transactions on Automatic Control*, 22:728–740, 1977.

[48] Y.-C. Ho and K.-C. Chu. Team decision theory and information structures in optimal control problems–Part I. *IEEE Transactions on Automatic Control*, 17:15–22, 1972.

[49] A. Isidori and C. I. Byrnes. Output regulation of nonlinear systems. *IEEE Transactions on Automatic Control*, 35:131–140, 1990.

[50] T. Kailath. *Linear Systems*. Prentice-Hall, 1980.

[51] R. Katoh and M. Mori. Control method of biped locomotion giving asymptotic stability of trajectory. *Automatica*, 20:405–414, 1984.

[52] J. D. Katzberg. Structured feedback control of discrete linear stochastic systems with quadratic costs. *IEEE Transactions on Automatic Control*, 22:232–236, 1977.

[53] D. E. Koditschek and E. Rimon. Robot navigation functions on manifolds with boundary. *Advances in Applied Mathematics*, 11:412–442, 1990.

[54] Y. Kuramoto. Self-entrainment of a population of coupled non-linear oscillators. In *International Symposium on Mathematical Problems in Theoretical Physics*, volume 39 of *Lecture Notes in Physics*, pages 420–422, 1975.

[55] J. Kurzweil and J. Jarnik. Limit processes in ordinary differential equations. *Journal of Applied Mathematics and Physics*, 38:241–256, 1987.

[56] H. Kwakernaak and R. Sivan. *Linear Optimal Control Systems*. Wiley, 1972.

[57] J. P. LaSalle. Some extensions of Liapunov's second method. *IRE Transactions on Circuit Theory*, 7:520–527, 1960.

[58] J. P. LaSalle. *The Stability of Dynamical Systems*. SIAM, 1976.

[59] A. M. Lyapunov. *The General Problem of the Stability of Motion*. PhD thesis, Kharkov Mathematical Society, 1892.

[60] D. Madjidian and L. Mirkin. Distributed control with low-rank coordination. *IEEE Transactions on Control of Network Systems*, 1:53–63, 2014.

[61] A.-R. Mansouri. Local asymptotic feedback stabilization to a submanifold: Topological conditions. *Systems & Control Letters*, 56:525–528, 2007.

[62] A.-R. Mansouri. Topological obstructions to submanifold stabilization. *IEEE Transactions on Automatic Control*, 55:1701–1703, 2010.

[63] R. Marino. High-gain feedback in non-linear control systems. *International Journal of Control*, 42:1369–1385, 1985.

[64] D. Q. Mayne. A canonical model for identification of multivariable linear systems. *IEEE Transactions on Automatic Control*, 17:728–729, 1972.

[65] A. Megretski and A. Rantzer. System analysis via integral quadratic constraints. *IEEE Transactions on Automatic Control*, 42:819–830, 1997.

[66] L. Moreau and D. Aeyels. Practical stability and stabilization. *IEEE Transactions on Automatic Control*, 45:1554–1558, 2000.

[67] M. Nagumo. Über die Lage der Integralkurven gewöhnlicher Differentialgleichungen. *Proceedings of the Physico-Mathematical Society of Japan*, 24:272–559, 1942.

[68] C. Nielsen and M. Maggiore. Output stabilization and maneuver regulation: A geometric approach. *Systems & Control Letters*, 55:418–427, 2006.

[69] C. Nielsen and M. Maggiore. On local transverse feedback linearization. *SIAM Journal on Control and Optimization*, 47:2227–2250, 2008.

[70] L. M. Pecora and T. L. Carroll. Synchronization in chaotic systems. *Physical Review Letters*, 64:821–825, 1990.

[71] H. Poincaré. *Les Méthodes Nouvelles de la Méchanique Celeste*. Gauthier-Villars, 1899.

[72] C. Pugh and M. Shub. Linearization of normally hyperbolic diffeomorphisms and flows. *Inventiones Mathematicae*, 10:187–198, 1970.

[73] K. Pyragas. Continuous control of chaos by self-controlling feedback. *Physics Letters A*, 170:421–428, 1992.

[74] X. Qi, M. V. Salapaka, P. G. Voulgaris, and M. Khammash. Structured optimal and robust control with multiple criteria: A convex solution. *IEEE Transactions on Automatic Control*, 49:1623–1640, 2004.

[75] M. Reyhanoglu, A. van der Schaft, N. H. McClamroch, and I. Kolmanovsky. Dynamics and control of a class of underactuated mechanical systems. *IEEE Transactions on Automatic Control*, 44:1663–1671, 1999.

[76] M. G. Safonov and V. V. Kulkarni. Zames-Falb multipliers for MIMO nonlinearities. *International Journal of Robust and Nonlinear Control*, 10:1025–1038, 2000.

[77] A. Sarlette, S. Bonnabel, and R. Sepulchre. Coordinated motion design on Lie groups. *IEEE Transactions on Automatic Control*, 55:1047–1058, 2010.

[78] A. Sarlette and R. Sepulchre. Consensus optimization on manifolds. *SIAM Journal on Control and Optimization*, 48:56–76, 2009.

[79] A. Sarlette, R. Sepulchre, and N. E. Leonard. Autonomous rigid body attitude synchronization. *Automatica*, 45:572–577, 2009.

[80] L. Scardovi, M. Arcak, and E. D. Sontag. Synchronization of interconnected systems with applications to biochemical networks: An input-output approach. *IEEE Transactions on Automatic Control*, 55:1367–1379, 2010.

[81] L. Scardovi, A. Sarlette, and R. Sepulchre. Synchronization and balancing on the *N*-torus. *Systems & Control Letters*, 56:335–341, 2007.

[82] C. W. Scherer. Structured finite-dimensional controller design by convex optimization. *Linear Algebra and its Applications*, 351-352:639–669, 2002.

[83] R. Sepulchre, M. Janković, and P. Kokotović. *Constructive Nonlinear Control*. Springer, 1997.

[84] R. Sepulchre, D. A. Paley, and N. E. Leonard. Stabilization of planar collective motion: All-to-all communication. *IEEE Transactions on Automatic Control*, 52:811–824, 2007.

[85] J. M. Soethoudt and H. L. Trentelman. The regular indefinite linear-quadratic problem with linear endpoint constraints. *Systems & Control Letters*, 12:23–31, 1989.

[86] T. Tanaka and C. Langbort. The bounded real lemma for internally positive systems and H-infinity structured static state feedback. *IEEE Transactions on Automatic Control*, 56:2218–2223, 2011.

[87] R. Tron, B. Afsari, and R. Vidal. Riemannian consensus for manifolds with bounded curvature. *IEEE Transactions on Automatic Control*, 58:921–934, 2013.

[88] V. I. Vorotnikov. *Partial Stability and Control*. Birkhäuser, 1998.

[89] P. Wieland, J. Wu, and F. Allgöwer. On synchronous steady states and internal models of diffusively coupled systems. *IEEE Transactions on Automatic Control*, 58:2591–2602, 2013.

[90] J. C. Willems. Almost invariant subspaces: An approach to high gain feedback design – part I: Almost controlled invariant subspaces. *IEEE Transactions on Automatic Control*, 26:235–252, 1981.

[91] J. C. Willems. Almost invariant subspaces: An approach to high gain feedback design – part II: Almost conditionally invariant subspaces. *IEEE Transactions on Automatic Control*, 27:1071–1085, 1982.

[92] J. C. Willems and C. Commault. Disturbance decoupling by measurement feedback with stability or pole placement. *SIAM Journal on Control and Optimization*, 19:490–504, 1981.

[93] F. W. Wilson. The structure of the level surfaces of a Lyapunov function. *Journal of Differential Equations*, 3:323–329, 1967.

[94] W. M. Wonham. *Linear Multivariable Control: A Geometric Approach*. Springer, 1979.

[95] W. M. Wonham and A. S. Morse. Decoupling and pole assignment in linear multivariable systems: A geometric approach. *SIAM Journal on Control*, 8:1–18, 1970.

[96] V. A. Yakubovich. Frequency-domain criteria for oscillation in nonlinear systems with one stationary nonlinear component. *Sibirskii Matematicheskii Zhurnal*, 14:1100–1129, 1973.

[97] G. Zames. On the input-output stability of time-varying nonlinear feedback systems part I: Conditions derived using concepts of loop gain, conicity, and positivity. *IEEE Transactions on Automatic Control*, 11:228–238, 1966.

[98] G. Zames. On the input-output stability of time-varying nonlinear feedback systems part II: Conditions involving circles in the frequency plane and sector nonlinearities. *IEEE Transactions on Automatic Control*, 11:465–476, 1966.

[99] G. Zames and P. L. Falb. Stability conditions for systems with monotone and slope-restricted nonlinearities. *SIAM Journal on Control*, 6:89–108, 1968.

[100] V. I. Zubov. *Methods of A. M. Lyapunov and their Applications*. Noordhoff, 1964.

Publications of the Author

[M1] W. Halter, J. M. Montenbruck, and F. Allgöwer. Geometric stability considerations of the ribosome flow model with pool. In *Proceedings of the 22nd International Symposium on Mathematical Theory of Networks and Systems*, 2016. to appear.

[M2] Y. Liu, J. M. Montenbruck, P. Stegagno, F. Allgöwer, and A. Zell. A robust nonlinear controller for nontrivial quadrotor maneuvers: Approach and verification. In *Proceedings of the 2015 IEEE / RSJ International Conference on Intelligent Robots and Systems*, pages 5410–5416, 2015.

[M3] J. M. Montenbruck and F. Allgöwer. Asymptotic stabilization of submanifolds embedded in Riemannian manifolds. *Automatica*. provisionally accepted for publication.

[M4] J. M. Montenbruck and F. Allgöwer. Structured state feedback controllers and optimal submanifold stabilization. *SIAM Journal on Control and Optimization*. submitted.

[M5] J. M. Montenbruck and F. Allgöwer. Pinning capital stock and gross investment rate in competing rationally managed firms. In *Proceedings of the 19th IFAC World Congress*, pages 10719–10724, 2014.

[M6] J. M. Montenbruck, M. Arcak, and F. Allgöwer. An input-output framework for submanifold stabilization. *IEEE Transactions on Automatic Control*. submitted.

[M7] J. M. Montenbruck, M. Arcak, and F. Allgöwer. Stabilizing submanifolds with passive input-output relations. In *Proceedings of the 54th IEEE Conference on Decision and Control*, pages 4381–4387, 2015.

[M8] J. M. Montenbruck, A. Birk, and F. Allgöwer. A convex conic underestimate of Laplacian spectra and its application to network synthesis. In *Proceedings of the 2015 European Control Conference*, pages 563–568, 2015.

[M9] J. M. Montenbruck and S. Brennan. Monitoring delayed systems using the Smith predictor. In *Proceedings of the 2012 American Control Conference*, pages 6235–6239, 2012.

[M10] J. M. Montenbruck, M. Bürger, and F. Allgöwer. Practical cluster synchronization of heterogeneous systems on graphs with acyclic topology. In *Proceedings of the 52nd IEEE Conference on Decision and Control*, pages 692–697, 2013.

[M11] J. M. Montenbruck, M. Bürger, and F. Allgöwer. Navigation and obstacle avoidance via backstepping for mechanical systems with drift in the closed loop. In *Proceedings of the 2015 American Control Conference*, pages 625–630, 2015.

[M12] J. M. Montenbruck, M. Bürger, and F. Allgöwer. Practical synchronization with diffusive couplings. *Automatica*, 53:235–243, 2015.

[M13] J. M. Montenbruck, M. Bürger, and F. Allgöwer. Synchronization of diffusively coupled systems on compact Riemannian manifolds in the presence of drift. *Systems & Control Letters*, 76:19–27, 2015.

[M14] J. M. Montenbruck, M. Bürger, and F. Allgöwer. Compensating drift vector fields with gradient vector fields for asymptotic submanifold stabilization. *IEEE Transactions on Automatic Control*, 61:388–399, 2016.

[M15] J. M. Montenbruck, H.-B. Dürr, C. Ebenbauer, and F. Allgöwer. Extremum seeking and obstacle avoidance on the special orthogonal group. In *Proceedings of the 19th IFAC World Congress*, pages 8229–8234, 2014.

[M16] J. M. Montenbruck, H.-B. Dürr, C. Ebenbauer, and F. Allgöwer. Extremum seeking with drift. In *Proceedings of the 1st IFAC Conference on Modelling, Identification and Control of Nonlinear Systems*, pages 126–130, 2015.

[M17] J. M. Montenbruck, G. S. Schmidt, A. Kecskeméthy, and F. Allgöwer. Two gradient-based control laws on SE(3) derived from distance functions. In A. Kecskeméthy and F. Geu Flores, editors, *Interdisciplinary Applications of Kinematics*, volume 26 of *Mechanisms and Machine Science*, pages 31–41. Springer, 2015.

[M18] J. M. Montenbruck, G. S. Schmidt, G. S. Seyboth, and F. Allgöwer. On the necessity of diffusive couplings in linear synchronization problems with quadratic cost. *IEEE Transactions on Automatic Control*, 60:3029–3034, 2015.

[M19] J. M. Montenbruck, G. S. Seyboth, and F. Allgöwer. Practical and robust synchronization of systems with additive linear uncertainties. In *Proceedings of the 2012 American Control Conference*, pages 6235–6239, 2012.

[M20] J. M. Montenbruck, D. Zelazo, and F. Allgöwer. Retraction balancing and formation control. In *Proceedings of the 54th IEEE Conference on Decision and Control*, pages 3645–3650, 2015.